The Econometric Analysis of Seasonal Time

In this book, Eric Ghysels and Denise R. Osborn provide a thorough and timely review of the recent developments in the econometric analysis of seasonal economic time series, summarizing a decade of theoretical advances in the area. The authors discuss the asymptotic distribution theory for linear stationary and nonstationary seasonal stochastic processes. They also cover the latest contributions to the theory and practice of seasonal adjustment, together with its implications for estimation and hypothesis testing. Moreover, a comprehensive analysis of periodic models is provided, including stationary and nonstationary cases. The book concludes with a discussion of some nonlinear seasonal and periodic models. The treatment is designed for an audience of researchers and advanced graduate students.

Eric Ghysels is the Edward M. Bernstein Professor of Economics and a Professor of Finance at the University of North Carolina at Chapel Hill. He has been a visiting professor or scholar at several major U.S., European, and Asian universities. He gave invited lectures at the 1990 World Congress of the Econometric Society, the 1995 American Statistical Association Meetings, the 1995 Brazilian Econometric Society Meetings, and the 1999 (EC)[2] Conference on financial econometrics. He has served on the editorial boards of the *Journal of the American Statistical Association,* the *Journal of Business and Economic Statistics,* the *Review of Economics and Statistics,* the *Journal of Empirical Finance,* and several Annals issues of the *Journal of Econometrics.* In 1999, he was Chair of the Business and Economics Statistics section of the American Statistical Association. He has published in the leading economics, finance, and statistics journals, and his main research interests are time series econometrics and finance.

Denise R. Osborn is the Robert Ottley Professor of Econometrics in the School of Economic Studies, University of Manchester, where she has taught since 1977. Professor Osborn was the inaugural chair of the Women's Committee of the Royal Economic Society (RES) and a past member of both the RES Executive and Council. She is currently Vice Chair of the Economics and Econometrics Panel for the 2001 Research Assessment Exercise in the U.K. and a member of the Training Board of the Economic and Social Research Council (ESRC). Her research interests have been primarily concerned with improving understanding of the dynamics of macroeconomic time series, recently focusing on seasonality and business cycles. Professor Osborn has published more than 30 papers in refereed journals, including *Journal of Econometrics, Journal of Applied Econometrics, Journal of Business and Economic Statistics, Journal of the American Statistical Association, Journal of the Royal Statistical Society (Series B),* and *Journal of Econometrics.*

Themes in Modern Econometrics

Managing Editor
PETER C. B. PHILLIPS, *Yale University*

Series Editors
ADRIAN PAGAN, *Australian National University*
CHRISTIAN GOURIEROUX, *CREST and CEPREMAP, Paris*
MICHAEL WICKENS, *University of York*

Themes in Modern Econometrics is designed to service the large and growing need for explicit teaching tools in econometrics. It will provide an organized sequence of textbooks in econometrics aimed squarely at the student population, and will be the first series in the discipline to have this as its express aim. Written at a level accessible to students with an introductory course in econometrics behind them, each book will address topics or themes that students and researchers encounter daily. While each book will be designed to stand alone as an authoritative survey in its own right, the distinct emphasis throughout will be on pedagogic excellence.

Titles in the Series

Statistics and Econometric Models: Volumes 1 and 2
CHRISTIAN GOURIEROUX and ALAIN MONFORT
Translated by QUANG VUONG

Time Series and Dynamic Models
CHRISTIAN GOURIEROUX and ALAIN MONFORT
Translated and edited by GIAMPIERO GALLO

Unit Roots, Cointegration, and Structural Change
G. S. MADDALA and IN-MOO KIM

Generalized Method of Moments Estimation
Edited by LÁSZLÓ MÁTYÁS

Nonparametric Econometrics
ADRIAN PAGAN and AMAN ULLAH

Econometrics of Qualitative Variables
CHRISTIAN GOURIEROUX
Translated by PAUL B. KLASSEN

Advance Praise for *The Econometric Analysis of Seasonal Time Series*

"Seasonal time series data are everywhere, and it is by now well understood that modeling such seasonality is important in various respects. This book gives an up-to-date account of what has been going on in the last 15 years in this area and hence what effectively is the current state of the art. The authors are to be commended for undertaking this enterprise, and I certainly believe it was worth the trouble. It simply is a good book!"

– Philip Hans Franses, *Erasmus University, Rotterdam*

"To many economists, seasonality is some sort of noise that is purged from the data before it is analyzed. Econometric research of the recent decades has revealed, however, that seasonal adjustment is a much more difficult exercise than had previously been assumed, as seasonal and non-seasonal data characteristics are linked. A full understanding of economic variables requires careful modeling of their seasonal features. Eric Ghysels and Denise Osborn have managed to survey a wide variety of statistical models of seasonality and of tests that are useful in discriminating among them. Their work will constitute valuable reference material for researchers and practitioners alike."

– Robert M. Kunst, *University of Vienna*

THE ECONOMETRIC ANALYSIS
OF SEASONAL TIME SERIES

ERIC GHYSELS
DENISE R. OSBORN

CAMBRIDGE
UNIVERSITY PRESS

PUBLISHED BY THE PRESS SYNDICATE OF THE UNIVERSITY OF CAMBRIDGE
The Pitt Building, Trumpington Street, Cambridge, United Kingdom

CAMBRIDGE UNIVERSITY PRESS
The Edinburgh Building, Cambridge CB2 2RU, UK
40 West 20th Street, New York, NY 10011-4211, USA
10 Stamford Road, Oakleigh, VIC 3166, Australia
Ruiz de Alarcón 13, 28014 Madrid, Spain
Dock House, The Waterfront, Cape Town 8001, South Africa

http://www.cambridge.org

© Eric Ghysels and Denise R. Osborn 2001

This book is in copyright. Subject to statutory exception
and to the provisions of relevant collective licensing agreements,
no reproduction of any part may take place without
the written permission of Cambridge University Press.

First published 2001

Printed in the United States of America

Typeface Times Roman 10/12 pt.   *System* LATEX 2_ε [TB]

A catalog record for this book is available from the British Library.

Library of Congress Cataloging in Publication Data
Ghysels, Eric, 1956–
    The econometric analysis of seasonal time series / Eric Ghysels, Denise R. Osborn.
      p. cm. – (Themes in modern econometrics)
    ISBN 0-521-56260-0 – ISBN 0-521-56588-X (pb)
    1. Econometrics. 2. Time-series analysis. I. Osborn, Denise R. II. Title. III. Series.
  HB139 .G49   2001
  330′.01′5195 – dc21   00-063070

ISBN 0 521 56260 0 hardback
ISBN 0 521 56588 X paperback

To my mother
E.G.

To Nicholas, Sarah, and Philip
D.R.O.

Contents

Foreword by Thomas J. Sargent

Seasonality means special annual dependence. Many weekly, monthly, or quarterly economic time series exhibit seasonality. Eric Ghysels and Denise Osborn's book is about specifying, econometrically testing, and distinguishing alternative forms of seasonality. This subject is of high interest to economists who use dynamic economic models to understand economic time series.

Dynamic economic theory seeks to describe and interpret economic time series in terms of the purposes and constraints of economic decision makers. Economic decision makers forecast the future to inform their decisions. Dynamic theory focuses on the relationship between forecasts and decisions, a relationship characterized by the Bellman equation of dynamic programming.

From the point of view of economic theory, the seasonal dependence of economic time series is especially interesting because substantial components of seasonal fluctuations are predictable. But in the most plausible models – those of indeterministic seasonality – seasonal fluctuations are not perfectly predictable. Optimizing behavior of decision makers combined with the predictability of seasonal time series leads to sharp, cross-equation restrictions between decision makers' rules and the seasonal time series whose forecasts impinge on their decisions. Particular statistical models of seasonality of the varieties described by Ghysels and Osborn differ in the detailed statistical structure of a time series of interest to a decision maker, and so they deliver different restrictions on decision makers' behavior and so on market prices and quantities.

That the predictability of seasonal fluctuations leads to sharp restrictions on decision rules has led to opposing responses from applied economists. One response has been to try to purge the time series of their seasonality – via seasonal adjustment procedures – and to focus analysis on the dynamics of the less predictable components of the time series. The opposite response has been to focus especially hard on the seasonal components, because they contain much detectable information about the preferences and constraints governing decisions. Ghysels and Osborn's book helps us to understand either approach, and to better implement both responses. As emphasized by Christopher Sims' work

xiv Foreword by Thomas J. Sargent

on estimating approximating models, a decision to use seasonally unadjusted data can be justified by a prior suspicion that one's model is least reliable for thinking about seasonal fluctuations.

Whether or not you want to use seasonally adjusted data, Ghysels and Osborn's book is of interest. Decisions to seasonally adjust data or to use raw data both require accurate specifications of the seasonal components of fluctuations. The strong seasonal fluctuations of economic time series give us a good laboratory for testing dynamic economic theories of asset pricing and accumulation of capital. Ghysels and Osborn's book will improve our use of that laboratory.

Palo Alto
November 1999

Preface

There has been a resurgence of interest in seasonality within economics in the past decade. Since the book by Hylleberg (1986) titled *Seasonality in Regression*, many developments have taken place. Some of these developments are covered in recent books by Franses (1996b) and Miron (1996). The former specializes on the subject of so-called periodic models, whereas the latter emphasizes economic aspects of seasonality. Also, surveys have been written by Franses (1996a), Ghysels (1994a), and Miron (1994), special issues of journals on the subject have been published by De Gooijer and Franses (1997) and Ghysels (1993), and a book of readings has been published by Hylleberg (1992).

The focus of this book is in some ways close to that of Hylleberg (1986), who gave a comprehensive description of the then-current state of the art. Our coverage will be more selective, however, as we mostly concentrate on the econometric developments of the past decade. Consequently, we will not discuss in detail subjects such as "The History of Seasonality," covered by Hylleberg (1986, Chapter 1) and Nerlove, Grether, and Carvalho (1995), or "The Definition of Seasonality," as in Hylleberg (1986, Chapter 2). Nor will we deal, at least directly, with the longstanding debate about the merits and pitfalls of seasonally adjusting series. It should also be made clear that this book focuses on theoretical developments and hence does not have the pretention to be empirical. It is not our purpose to analyze real data, beyond the presentation of some series that illustrate different types of seasonality. Similarly, we give no advice to the practitioner who wishes to seasonally adjust on how this adjustment should be performed. Rather, we aim to present a coherent account of the current state of the econometric theory for analyzing seasonal time series processes.

Important developments have taken place over the past decade. Parallel to the advances in the econometrics of unit root nonstationarity, we have witnessed corresponding progress in the analysis of seasonal nonstationarity. In Chapter 3 we provide a comprehensive coverage of unit root nonstationarity in a seasonal context. Chapter 2 precedes this analysis and covers the more traditional topic of deterministic seasonal processes, although here too there has been substantial progress in the 1990s. Although we are not concerned with the pros and cons of seasonal adjustment, it is a matter of fact that seasonality is frequently modeled implicitly through the use of seasonal adjustment filters. The U.S. Bureau of the Census and Agustin Marvall at the Bank of Spain have both recently developed new procedures for seasonal adjustment, namely the X-12 and SEATS/TRAMO programs, respectively. Chapter 4 is almost entirely devoted to these procedures. Since the seminal papers by Sims (1974) and Wallis (1974), much more has been learned about estimation with filtered data, which is why an entire chapter (Chapter 5) is devoted to it. Periodic models of seasonality were introduced into the econometrics literature during the late 1980s; Chapter 6 summarizes this literature and its subsequent progress.

Traditionally, the topic of seasonality is put in a context of macroeconomic time series sampled at a monthly or quarterly frequency. In one area, however, namely financial econometrics, new high-frequency data sets have become available with series sampled on a transaction basis. The availability of such series has created considerable interest in models for so-called intraday seasonality. This is one of the topics of Chapter 7, together with other topics related to the nonlinear analysis of seasonal time series.

Chapter 1 is a guided tour of the substantive material that follows. We start with a set of examples of seasonal economic and financial time series to motivate the discussion. The examples considered include some empirical series and theoretical models. The analysis in this first chapter is not meant to be rigorous, but rather illustrative, since the main purpose is to provide the reader with a general overview of the different types of seasonal processes considered in later chapters.

We have written this book for an audience of researchers and graduate students familiar with time series analysis at the level of, for instance, Hamilton (1994a). Thus we do not review certain basic results of the econometrics of stationary and nonstationary time series. For example, we refer to I(0) and I(1) processes without reviewing their basic properties. Also, we use spectral analysis, assuming familiarity with spectral domain techniques. There are many other cases in which we assume that the reader has sufficient background knowledge to understand the concepts and methods. We make no apology for this since there are good recent textbooks, including Hamilton (1994a), which cover this material.

This book contains many results that have emerged from joint work undertaken by the authors with others. We have had the privilege and pleasure to work with many co-authors whose intellectual input we would like to acknowledge. Material in this book either directly or indirectly covers joint work of the authors with Chris Birchenhall, Tim Bollerslev, Alice Chui, Clive Granger, Alastair Hall, Saeed Heravi, Hahn Lee, Offer Lieberman, Jason Noh, Pierre

Perron, Paulo Rodrigues, Pierre Siklos, and Jeremy Smith. We would also like to thank Robert Taylor for reading and commenting on some of the material and Paulo Rodrigues for assistance with the figures. Peter Phillips, Patrick McCartan, and Scott Parris provided invaluable guidance and help while we were writing this monograph, and Nicholas Berman patiently proofread the entire manuscript. We also benefited from insightful comments of three referees who read preliminary drafts of the book. Finally, we owe an unmeasurable debt to Sandi Lucas, who keyboarded various drafts of the book with devotion beyond the call of duty.

Eric Ghysels
Chapel Hill, U.S.A.

Denise R. Osborn
Manchester, U.K.

List of Symbols and Notation

D_s	periodic first difference operator for season s;
$f_y(w)$	spectral density of y at frequency ω;
$I(d)$	process integrated of order d;
L	lag operator, $Ly_t = y_{t-1}$;
l_k	$k \times 1$ vector of ones (k may be omitted);
m_s	seasonal mean shifts, $s = 1, \ldots, S$;
$PI(d)$	process periodically integrated of order d;
S	number of seasons per year;
s_t	season in which t falls, $s_t = 1 + [(t-1) \bmod \quad S]$;
$SI(d)$	process seasonally integrated of order d;
T	number of observations in sample;
T_τ	number of years of sample observations;
U_t	vector i.i.d. disturbance, $U_t = (\varepsilon_{1t}, \ldots, \varepsilon_{Nt})'$;
$W_i(r)$	standard Wiener process;
Y_t	generic vector time series, $Y_t = (y_{1t}, \ldots, y_{Nt})'$;
Y_τ	vector time series of the S seasons of year τ, $Y_\tau = (y_{1\tau}, \ldots, y_{S\tau})'$, $\tau = 1, \ldots, T_\tau$ (similarly, U_τ);
$y_{s\tau}$	observation on y_t relating to season s of year τ ($s = 1, \ldots, S$, $\tau = 1, \ldots, T_\tau$) (similarly, $\varepsilon_{s\tau}$);
y_t	generic univariate time series observed over $t = 1, \ldots, T$;
$y_t^{(1)}/y_t^{(2)}/y_t^{(3)}$	linear filtered series of y_t, used in HEGY procedure;
y_t^h	holiday component of y_t;
y_t^i	irregular component of y_t;
y_t^{ns}	nonseasonal component of y_t;

y_t^s	seasonal component of y_t;
y_t^{tc}	trend cycle component of y_t;
y_t^{td}	trading-day component of y_t;
y_t^{X11}/y_t^{LSA}	X11/linear seasonally adjusted filtered series of y_t;
z_t	weakly dependent univariate process.

Greek

α_S	coefficient tested in DHF t test;
γ	drift;
Δ_S	seasonal (annual) difference operator, $\Delta_S = 1 - L^S$;
δ_{st}	seasonal dummy variable for observation t and season s, $\delta_{st} = 1$; if $s = 1 + [(t - 1) \bmod \quad S]$ and 0 otherwise;
ε_t	univariate i.i.d. disturbance;
$\Theta(L)$	vector moving average operator, $\Theta(L) = \Theta_0 - \Theta_1 L - \cdots - \Theta_q L^q$;
$\theta(L)$	univariate moving average operator, $\theta(L) = 1 - \theta_1 L \cdots - \theta_q L^q$;
$\theta_S(L^S)$	seasonal polynomial MA lag operator;
π_i	coefficient tested in HEGY test;
$\rho(k)$	autocorrelation function at lag k;
$\tilde{\sigma}$	degrees of freedom corrected estimator of σ;
τ_t	year of observation t, $\tau_t = 1 + \text{int}[(t - 1)/S]$;
$\Phi(L)$	vector autoregressive operator, $\Phi(L) = \Phi_0 - \Phi_1 L - \cdots - \Phi_p L^p$;
$\phi(L)$	univariate autoregressive operator, $\phi(L) = 1 - \phi_1 L \cdots - \phi_p L^p$;
ϕ_{ij}^P	periodic autoregressive coefficient for variable i at lag j (j may be omitted for periodic AR of order 1; P may be omitted when periodic context is clear);
$\phi_S(L^S)$	seasonal polynomial AR lag operator.

Processes

ARMA process for seasonal component of y_t	$\phi_s(L)y_t^s = \theta_s(L)\varepsilon_t^s$ (similarly, other components);

k_i	coefficient of cointegrating relationship;
Regression model	$y_t = x_t\beta + u_t$;
Univariate ARMA process	$\phi(L)y_t = \theta(L)\varepsilon_t$;
Vector ARMA process	$\Phi(L)Y_t = \Theta(L)U_t$.

1 Introduction to Seasonal Processes

1.1 Some Illustrative Seasonal Economic Time Series

Economic time series are usually recorded at some fixed time interval. Most macroeconomic aggregate series are released on a monthly or quarterly frequency. Other economic data, particularly financial series, are available more frequently. In fact, some financial market data are available on a transaction basis and hence are no longer sampled at fixed intervals. Throughout this book we will focus almost exclusively on data sampled at a monthly or quarterly frequency. There is an emerging literature dealing with data sampled at higher frequencies, such as weekly, daily, or intradaily, possibly unequally spaced. This literature is still in its infancy, particularly compared with what we know about quarterly or monthly sampled data.

The purpose of this section is to provide some illustrative examples of seasonal time series of interest within economics. It is certainly not our intention to be exhaustive. Instead, we look at several series that reveal particularly interesting features and stylized facts pertaining to seasonality. We also discuss various ways to display seasonal series.

Figure 1.1 has four panels displaying U.K. and U.S. monthly growth rates in industrial production over roughly 35 years. The left panels represent unadjusted raw series, whereas the right panels show the corresponding seasonally adjusted series. Both the U.K. and U.S. raw series show very regular seasonal patterns that are clearly dominant and seem to obscure and dwarf the non-seasonal fluctuations. The nonseasonal fluctuations are filtered from the raw data and appear in the right panels (note the difference in scale to appraise the dominance of the seasonal fluctuations). The seasonally adjusted series are the most often used; that is; they are typically released to the general public, appear in the financial press, and so on. There has been a long debate about the merits and dangers of seasonally adjusting economic time series, a debate that will not be covered in this book. Lucid discussions appear in Ghysels (1994a), Granger (1978), Hylleberg, Jørgensen, and Sørensen (1993),

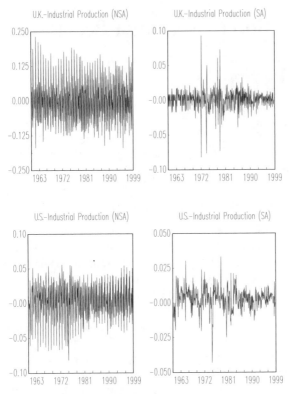

Figure 1.1. U.K. and U.S. industrial production – monthly NSA (not seasonally adjusted) and SA (seasonally adjusted).

and Miron (1994) among many others. Obviously, to obtain a seasonally adjusted series we need to subtract from (or divide) the raw series (by) an estimate of the seasonal component. Whether this component is assumed orthogonal to the nonseasonal one, and therefore void of interest to economists, and whether the seasonal is best viewed as deterministic or not are some of the basic questions being debated. While we do not elaborate explicitly on the pros and cons of seasonal adjustment, we do cover quite extensively some of the basic questions underlying this debate. For example, the distinct features of deterministic and stochastic models of seasonality are discussed in great detail in Chapters 2 and 3.

We can infer from Figure 1.1 that the seasonal patterns in the U.K. and U.S. are not quite the same. U.K. industrial production shows far greater variability and, taking into account the differences in scale, certainly larger peaks. It is clear from the right panels, however, that the greater variability of U.K. series is partly due to the nonseasonal fluctuations.

Two series appear in Figure 1.2; these are U.K. retail sales of clothing (shown as the first difference of the logarithm, in the top two panels) and U.K. retail credit: net lending (lower two panels). Retail sales of clothing show an extremely regular seasonal pattern, except perhaps that the third-quarter dip appears more pronounced toward the end of the sample. The top left plot is a standard time series plot. The top right graph represents the same series plotted as four curves, each corresponding to a specific quarter of each calendar year. Since clothing is a basic consumption good, it is not surprising to find hardly any business cycle movements in the series. The time series plot of annual quarters reveals that a simple model with a distinct mean for each quarter probably goes a long way toward describing the behavior of the series. Plotting each quarter as a separate curve is a particularly useful tool for examining the seasonal behavior of a series. Such plots can be found in Hylleberg (1986), and the construction of the underlying series, that is, a multivariate vector process of quarterly observations sampled annually, forms the basic building block of several test statistics [e.g., Franses (1994) and Canova and Hansen (1995), among others] that have been proposed recently and will be discussed later.

While the retail sales of clothes is highly seasonal and regular, it is clear from the graphs appearing in the lower two panels of Figure 1.2 that net lending of retail credit is also highly seasonal but clearly does not feature stable seasonal patterns. The time series plot shows fluctuations of increasing amplitude. The plot of quarterly curves drift apart and cross each other. The two examples in Figure 1.2 are almost perfect polar cases. Sales of clothing is a relatively easy series to model in comparison to net lending. One obvious implication is that seasonal adjustment of net lending is a rather daunting task. Of the two series, it is ironically also the one in which the nonseasonal component is more interesting.

When we restrict our attention to macroeconomic time series, then the class of seasonal models is confined to processes with dynamic properties at periods of a quarter or a month. However, when financial time series are studied, then our interest shifts to seasonal patterns at the daily level together with seasonal properties in higher moments. The last example, which appears in Figure 1.3, illustrates this. It shows the daily returns and daily returns squared of the Standard and Poor (S & P) 500 stock index as well as its monthly and daily averages. It is clear from Figure 1.3 that distinct patterns appear both in the mean and volatility at the daily and monthly level.[1]

In summary, our selection of empirical series has illustrated that seasonality has many different manifestations. It is no surprise, therefore, that a rich toolkit of econometric techniques has been developed to model seasonality.

---

[1] One of the early studies on daily returns patterns is Gibbons and Hess (1981). In Chapter 7, we provide further details.

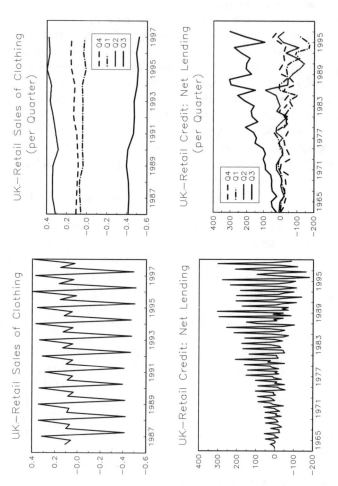

Figure 1.2. U.K. retail sales of clothing and retail credit.

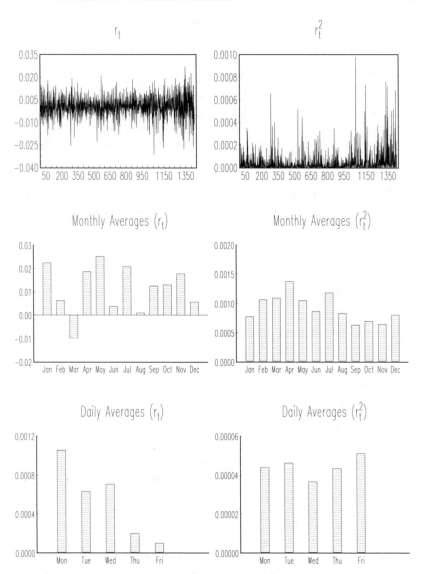

Figure 1.3. Returns and squared returns of S&P 500 index.

1.2 Seasonality in the Mean

In this section, we review a class of time series processes that model seasonal mean behavior. This includes most of the standard processes, such as deterministic seasonal mean shifts, stochastic stationary and nonstationary processes, and unobserved components ARIMA models. A subsection will be devoted to each class of process.

1.2.1 Deterministic Seasonality

From knowledge of the season in which the initial observation falls, we can deduce the season for all subsequent values of t. For simplicity of exposition, we assume that $t = 1$ corresponds to the first season of a year (that is, the first quarter for quarterly data or January for monthly observations), and we denote the season in which observation t falls as s_t. With S observations per year, s_t can be obtained mathematically as $s_t = 1 + [(t-1) \bmod S]$. (That is, s_t is one plus the integer remainder obtained when $t - 1$ is divided by S.) It will also be useful to have a notation for the year in which a specific observation falls, and we refer to this as τ_t. This can be found as $\tau_t = 1 + \text{int}\,[(t-1)/S]$, where int denotes the integer part.

Let us consider a univariate process y_t, which has the following representation:

$$y_t = \sum_{s=1}^{S} \delta_{st} m_s + z_t, \qquad t = 1, \dots, T, \tag{1.1}$$

where δ_{st} is a seasonal dummy variable that takes the value one in season s (more formally $\delta_{st} = 1$ if $s_t = s$ for $s = 1, \dots, S$) and is zero otherwise. The process z_t is assumed to be a weakly stationary zero mean process.[2] The first term on the right-hand side of (1.1) represents deterministic level shifts, while the second term z_t is stochastic and may even exhibit seasonal features as will be discussed in the next subsection. Hence, for any given season s the unconditional mean of y_t is m_s. In Figure 1.4, Panel A, we display a time series plot of sample size $T = 40$ of a deterministic seasonal process with z_t i.i.d. $N(0, 1)$ and $S = 4$ with $m_1 = -1.5$, $m_2 = -0.5$, $m_3 = 0.5$, and $m_4 = 1.5$; DGP stands for data-generating process. Figure 1.4 has four panels, similar to those used for the retail sales of clothing and net lending time series in the previous section. Namely, Panel B of Figure 1.4 displays the data as four annual time series plots of quarters. Panels C and D repeat this exercise for a large sample of 50 years ($T = 200$) instead of 10 years. It is clear from this simulated series example that the four curves of quarterly data appearing in panels B and D can cross and show considerable variability, unlike the U.K. retail sales of clothing series. In large samples (Panel D) the quarterly curves do not drift apart, however, since the process features mean reversion.

---

[2] For a definition of weakly stationary or covariance stationary processes, see, e.g., Hamilton (1994a).

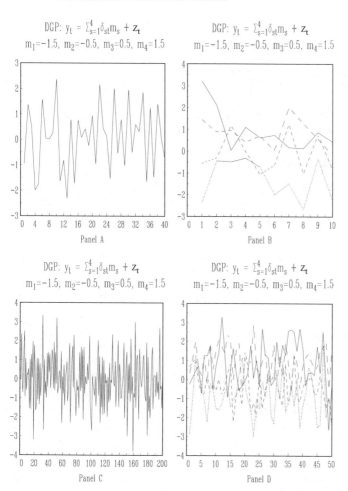

Figure 1.4. Two examples of a deterministic seasonal process.

A deterministic function with period S can also be equivalently written in terms of sines and cosines, namely

$$\sum_{s=1}^{S} \delta_{st} m_s = \sum_{k=1}^{S/2} \left[\alpha_k \cos \left(\frac{2\pi kt}{S} \right) + \beta_k \sin \left(\frac{2\pi tk}{S} \right) \right] \qquad (1.2)$$

for $t = 1, \ldots, T$, where

$$\alpha_k = \frac{2}{S} \sum_{s=1}^{S} m_s \cos \left(\frac{2\pi kj}{S} \right), \qquad k = 1, 2, \ldots, \frac{S}{2} - 1, \qquad (1.3)$$

$$\alpha_{S/2} = \frac{1}{S} \sum_{s=1}^{S} m_s \cos(\pi j), \tag{1.4}$$

$$\beta_k = \frac{2}{S} \sum_{s=1}^{S} m_s \sin\left(\frac{2\pi kj}{S}\right), \qquad k = 1, 2, \ldots, \frac{S}{2}. \tag{1.5}$$

This trignometric representation is useful in separating seasonality from the overall mean of the process, as discussed in Chapter 2.

1.2.2 Linear Stationary Seasonal Processes

The class of autoregressive-moving average (ARMA) processes is widely used to describe univariate time series. Some ARMA processes feature stochastic seasonality. A prominent example is the first-order seasonal autoregressive process, defined as

$$y_t = \phi_S y_{t-S} + \varepsilon_t, \tag{1.6}$$

where ε_t is i.i.d. zero mean and variance σ_ε^2, henceforth denoted $\varepsilon_t \sim$ i.i.d. $(0, \sigma_\varepsilon^2)$, and $|\phi_S| < 1$. Using the lag operator, that is, $L^k y_t = y_{t-k}$, we can write (1.6) as

$$(1 - \phi_S L^S) y_t = \varepsilon_t. \tag{1.7}$$

As there is no intercept, the unconditional mean of the process regardless of the season is equal to zero. This property is different from the mean property of the deterministic seasonal process. However, the mean conditional on past y_t exhibits seasonal patterns, since

$$E(y_t \mid y_{t-1}, \ldots) = \phi_S y_{t-S}. \tag{1.8}$$

Moreover, the autocorrelation function takes on the values

$$\rho(kS) = \phi_S^k, \qquad k = 1, 2, \ldots, \tag{1.9}$$

and $\rho(k) = 0$ for all other k. As a result of property (1.9), autocorrelation is exhibited at seasonal lags only and this nonzero autocorrelation diminishes in magnitude over time. Consequently, the only pattern in y_t is associated with seasonality, but this seasonality is transitory in the sense that the series is mean reverting toward its expected value of zero, irrespective of the season.

The idea of a seasonal autoregressive (AR) process of order one can be extended to higher-order processes possibly with moving average (MA) seasonal polynomials, namely

$$\phi_S(L^S) y_t = \theta_S(L^S) \varepsilon_t, \tag{1.10}$$

where $\phi_S(L^S) \equiv 1 - \sum_{i=1}^{P} \phi_{iS} L^{iS}$ and $\theta_S(L^S) \equiv 1 - \sum_{i=1}^{Q} \theta_{iS} L^{iS}$, where both polynomials have their roots outside the unit circle and have no common

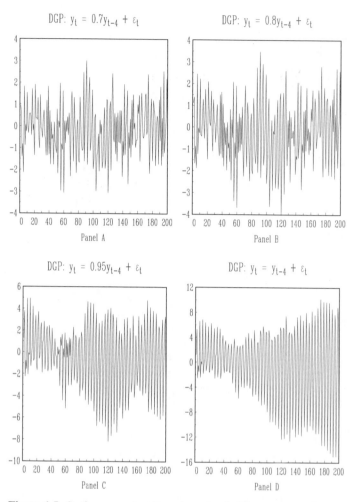

Figure 1.5. Stationary and unit root seasonal AR processes of order one: 50-year samples of quarterly data.

roots. Such processes are generally called seasonal ARMA processes, and their autocorrelation function appears in Box and Jenkins (1976) or in Fuller (1996).

To appraise the features of this class of processes, let us turn to Figure 1.5, which has four panels. Panels A–C display 50-year random samples drawn from a quarterly seasonal autoregressive process of order one with values of $\phi_4 = 0.7, 0.8$, and 0.95. Obviously, as ϕ_4 increases the process appears to be less mean reverting. Panel D in Figure 1.5 pertains to a seasonal unit root process, which will be discussed in the next subsection. The same processes are repeated

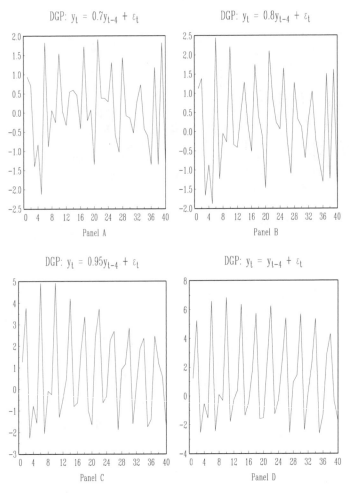

Figure 1.6. Stationary and unit root seasonal AR processes of order one: 10-year samples of quarterly data.

in Figure 1.6 for a much smaller sample size, namely 10 years. Note how much more similar the four processes appear to be in this latter case.

An alternative representation for seasonal ARMA processes, or in fact all ARMA processes following the spectral representation theorem [see Priestley (1981)], involves Fourier transforms. Let $f_y(\omega)$ be the spectral density of y at frequency ω; then

$$f_y(\omega) = \frac{|\theta_S(e^{-i\omega S})|^2}{|\phi_S(e^{-i\omega S})|^2}\sigma_\varepsilon^2 \quad -\pi < \omega < \pi, \tag{1.11}$$

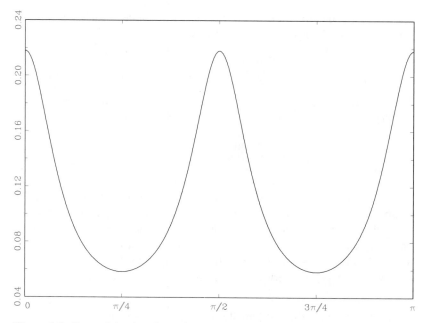

Figure 1.7. Spectral density of a stationary quarterly seasonal AR process of order one.

where $|\theta_S(e^{-i\omega S})|^2 = \theta_S(e^{-i\omega S})\theta_S(e^{i\omega S})$ and $\theta_S(e^{-i\omega S}) \equiv 1 - \sum_{k=1}^{P} \theta_{kS} e^{-i\omega kS}$, with an analogous definition of $|\phi_S(e^{-i\omega S})|^2$. It can be shown that for the special case of a stationary seasonal AR process of order one appearing in (1.6), the spectral density becomes

$$f_y(\omega) = \sigma_\varepsilon^2 / \left[1 + \phi_S^2 - 2\phi_S \cos(S\omega) \right]. \tag{1.12}$$

Figure 1.7 displays the spectral plot of the quarterly process with $\phi_4 = 0.8$.

1.2.3 Nonstationary Unit Root Processes

Economic time series typically exhibit unit root nonstationarity. This feature may also be present in the seasonal behavior. The simplest process that exhibits those features is the so-called seasonal random walk:

$$\Delta_S y_t = \varepsilon_t, \tag{1.13}$$

where $\Delta_S \equiv (1 - L^S)$ is the seasonal differencing operator. Obviously, this process can be obtained from (1.6) with $\phi_S = 1$. It is therefore called a seasonal unit root process. The spectral density of this process has infinite peaks at the so-called seasonal frequencies $\omega = 2\pi/S$ [see e.g., Priestley (1981)]. As in the stationary case and with the simplifying assumption $y_{-s+1} = \cdots = y_0 = 0$, there is no seasonality in the mean, whereas (1.8) applies with $\phi_S = 1$. In this case the seasonality is persistent in the sense that

$$E(y_t \mid y_{t-1}, \ldots) = E(y_{t+kS} \mid y_{t-1}, \ldots)$$
$$= y_{t-S}, \qquad k = 1, 2, \ldots.$$

Thus, the process does not exhibit mean reversion. Examples of seasonal random walk processes appear in Panel D in Figures 1.5 and 1.6.

1.2.4 Unobserved Component Models

Seasonality is often viewed as just one of several components of observed economic time series. There is a long tradition of viewing trends, cycles, seasonality, and irregularity as separately generated components. This idea is deeply entrenched in the growth, business cycle, and seasonality literatures and is, of course, the basis of seasonal adjustment. Observed time series are assumed to be a function of these components, none of which is separately observed. A linear unobserved component model, in particular, is

$$y_t = y_t^{tr} + y_t^c + y_t^s + y_t^i, \tag{1.14}$$

where the components are mutually independent. The superscripts in (1.14) indicate the trend, business cycle, seasonal and irregular components. It should come as no surprise that the nature of the components is a fundamental issue on which there is no general agreement. However, in some special cases there are straightforward decompositions. For instance, an unobserved component model of a seasonal random walk has the following form:

$$(1 - L)\,y_t^{ns} = \varepsilon_t^{ns}, \tag{1.15}$$

$$(1 + L + \cdots + L^{S-1})y_t^s = \varepsilon_t^s, \tag{1.16}$$

where y_t^{ns} is the nonseasonal component of y_t comprising of y_t^{ts}, y_t^c, and y_t^i. The first equation, (1.15), represents a nonseasonal random walk and captures the zero-frequency component of the seasonal random walk. The factorization of Δ_S into $(1 - L)$ and $(1 + L + \cdots + L^{S-1})$ is quite common and is used in tests of seasonal unit roots, as further discussed in Chapter 3.

There is in general no unique decomposition of y_t into separate orthogonal components; see, for instance, Bell and Hillmer (1984). Seasonal adjustment of economic time series has to be based on a particular decomposition; for more on this, see Chapter 4. The specific assumptions one makes to carry out such adjustments will determine the final properties of the estimated nonseasonal component. In a general context there are various classes of models that have been suggested. For instance, unobserved component integrated ARMA (ARIMA) models specify particular ARIMA structures for y_t^{ns} and y_t^s; see, for instance, Bell and Hillmer (1984), Maravall (1995), Nerlove et al. (1995), and others. State space models are often an alternative vehicle to formulate unobserved component models; see Harvey (1993), among others.

1.3 Periodic Processes

Deterministic seasonality, discussed in Subsection 1.2.1, was modeled by means of seasonally varying intercepts. The stochastic seasonal models feature fixed parameters with seasonal lags in the AR and/or MA polynomials. Periodic models feature seasonally varying parameters in ARMA-type models. The simplest case is an autoregressive process without intercept, namely

$$y_t = \sum_{s=1}^{S} \delta_{st} \phi_s^P y_{t-1} + \varepsilon_t, \tag{1.17}$$

where the superscript P stands for periodic, indicating the seasonal variation in the parameters. As there is no intercept, the unconditional mean of y_t is always zero. Yet, the conditional mean can be written as

$$E(y_t \mid y_{t-1}, \ldots) = \sum_{s=1}^{S} \delta_{st} \phi_s^P y_{t-1}.$$

In a conventional sense this process is not stationary, as the lag one auto-correlations are seasonally varying. There is, however, a periodic version of stationarity. One easy way to see this is to take $S = 2$ and construct the following bivariate representation:

$$\begin{pmatrix} 1 & 0 \\ -\phi_2^P & 1 \end{pmatrix} \begin{pmatrix} y_{1\tau} \\ y_{2\tau} \end{pmatrix} = \begin{pmatrix} 0 & \phi_1^P \\ 0 & 0 \end{pmatrix} \begin{pmatrix} y_{1\tau-1} \\ y_{2\tau-1} \end{pmatrix} + \begin{pmatrix} u_{1\tau} \\ u_{2\tau} \end{pmatrix}, \tag{1.18}$$

where $\tau = \text{int}[(t+1)/2]$ and $y_{s\tau} = y_t$ with $t = 2(\tau - 1) + s$ for $s = 1$ and 2. Hence in the context of an annual seasonal, $y_{1\tau}$ and $y_{2\tau}$ represent the first and second semesters, respectively, of year τ, while $u_{1\tau}$ and $u_{2\tau}$ represent the corresponding disturbances. Therefore the bivariate system of equations in (1.18) involves annual vectors of stacked semesters. This system can be viewed as stationary under suitable regularity conditions [see, e.g., Lütkepohl (1991)]. In this particular case this condition amounts to $|\phi_1^P \phi_2^P| < 1$.

Figures 1.8 and 1.9 display two processes; the first takes values $\phi_1^P = 0.5$ and $\phi_2^P = 1.1$. Hence, the process has two seasons. Panels A and B of Figure 1.8 display samples of sizes $T = 40$ and 200 plotted as annual series, with each curve representing a season. Panels A and B of Figure 1.9 redisplay the series as regular time series plots.

One nontrivial consequence of the vector representation appearing in (1.18) is that periodic processes not only have seasonal variation in the autocorrelation function, but also in the unconditional variances; that is, there is seasonal

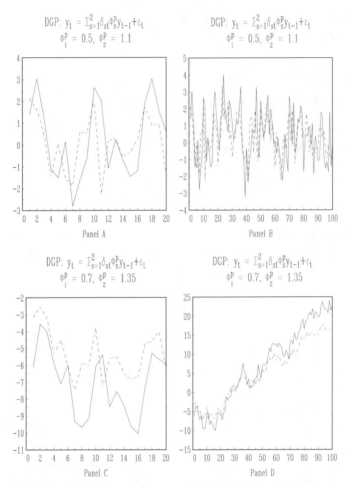

Figure 1.8. Periodic AR processes of order one: 20- and 100-year samples plotted as annual time series.

heteroskedasticity in y_t. To establish this, it suffices to invert the matrix on the left-hand side (1.18) and compute the unconditional variances from that. This will be discussed in further detail in Chapter 6. Further generalizations of the periodic AR process of order one to higher-order periodic ARMA processes are covered in the same chapter.

When the product $|\phi_1^P \phi_2^P| = 1$ in (1.18), the process is called periodically integrated; see Osborn et al. (1988). A single unit root is present when $\phi_1^P \phi_2^P = 1$ and the vector representation is then cointegrated; see, for instance, Franses (1994) and Osborn (1993). To highlight this feature we display two random

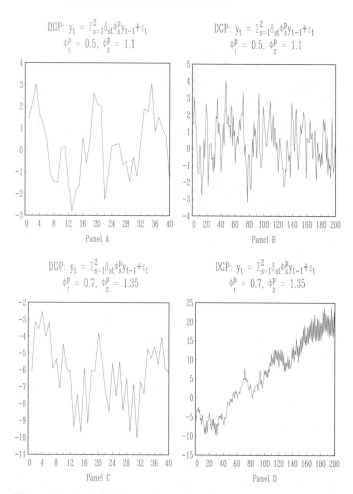

Figure 1.9. Periodic AR processes of order one: time series plots of 20- and 100-year samples.

draws of a periodic autoregressive process of order one with $\phi_1^P = 0.7$ and $\phi_2^P = 1.35$. The product of these two parameters is 0.945; hence it approaches the periodically integrated case. Panels C and D of Figures 1.8 and 1.9 display this process. Particularly, Panel D shows the strong persistence in the series. In Panel D of Figure 1.8, we also get a hint of cointegration between the time series plots of $y_{1\tau}$ and $y_{2\tau}$. Other forms of nonstationarity can also occur in higher-order systems; for instance, a process can be periodic in first differences, in which case it is a periodic $I(1)$ process. See further details in Chapter 6.

1.4   Seasonality in Higher Moments

In Chapter 7, we will explore some classes of nonlinear time series processes that feature seasonality or periodicity in higher moments. The leading examples will be processes that feature seasonal and periodic conditional heteroskedasticity.

Chapter 7 will be far more reflective than the previous chapters. There are many nonlinear time series models, and for each type of model one can typically formulate a seasonal extension. Many forms of nonlinearity are difficult to identify, particularly in relatively small samples. Having a seasonal dimension to most models makes the demands on the data usually even worse, because, say, in a threshold model there will typically be not one threshold but a seasonally varying threshold. Likewise, in a Markov switching model, there is not a single transition matrix but potentially one for every season. Financial time series are probably the most suitable to fit nonlinear models, primarily because of two reasons: (1) abundance of data and (2) most of the interesting dynamics are in the higher-order moments.

1.4.1   Stochastic Seasonal Unit Roots

The analysis in Chapter 7 focuses almost exclusively on seasonal conditional heteroskedasticity. The first class of models is closely related to the first-order seasonal autoregressive processes appearing in (1.6), and in particular the seasonal unit root process appearing in (1.13). Instead of assuming a fixed autoregressive parameter, let us consider a random autoregressive coefficient centered around unity. More precisely,

$$y_t = (1 + \widetilde{\phi}_t)y_{t-s} + \varepsilon_t, \tag{1.19}$$

where $E(\widetilde{\phi}_t) = 0$ and $\widetilde{\phi}_t = \rho\widetilde{\phi}_{t-s} + \xi_t$ with ξ_t i.i.d. $N(0, \sigma_\xi^2)$, whereas ε_t is i.i.d. $N(0, \sigma_\varepsilon^2)$. This process is obviously closely related to the class of linear time series involving seasonal unit roots. Yet, these so-called stochastic unit processes [see Granger and Swanson (1997), Leybourne, McCabe, and Tremayne (1996), McCabe and Tremayne (1995), and Taylor and Smith (1998)] feature conditional heteroskedasticity in the seasonally differenced series.

1.4.2   Seasonal GARCH

An alternative and more direct way of modeling seasonal heteroskedasticity involves processes of the autoregressive-conditional heteroskedasticity (ARCH) type, which derive from the original paper of Engle (1982). A simple example of such processes is the generalized ARCH (GARCH) process

$$\varepsilon_t^2 = \mu + \phi\varepsilon_{t-s}^2 + v_t + \theta v_{t-s}, \tag{1.20}$$

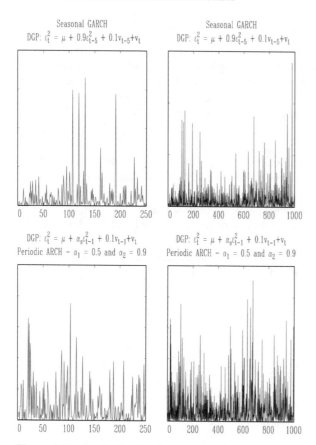

Seasonal GARCH

DGP: $\varepsilon_t^2 = \mu + 0.9\varepsilon_{t-5}^2 + 0.1v_{t-5}+v_t$

Seasonal GARCH

DGP: $\varepsilon_t^2 = \mu + 0.9\varepsilon_{t-5}^2 + 0.1v_{t-5}+v_t$

DGP: $\varepsilon_t^2 = \mu + \alpha_s\varepsilon_{t-1}^2 + 0.1v_{t-1}+v_t$

Periodic ARCH – $\alpha_1 = 0.5$ and $\alpha_2 = 0.9$

DGP: $\varepsilon_t^2 = \mu + \alpha_s\varepsilon_{t-1}^2 + 0.1v_{t-1}+v_t$

Periodic ARCH – $\alpha_1 = 0.5$ and $\alpha_2 = 0.9$

Figure 1.10. Seasonal and periodic GARCH processes.

where $y_t = \varepsilon_t$ and v_t is i.i.d. $N(0, 1)$. The top two panels of Figure 1.10 display a small sample (250 observations) and a larger sample (1000 observations) of a seasonal GARCH(1,1) process with $\phi = 0.9$ and $\theta = 0.1$, where $S = 5$. These could be thought of as representing a weekly seasonal cycle of daily observations (hence the small sample of 250 observations is roughly a year's worth of data). We observe the volatility clustering that is often seen in financial time series and that ARCH-type models attempt to mimic.

Seasonal GARCH processes exhibit many features similar to standard GARCH, such as fat-tailed unconditional distributions and other properties that will be discussed in Chapter 7. They can also produce different means for the volatility (like squared returns in the lower right panel of Figure 1.3) for each day of the week.

1.4.3 Periodic GARCH

Since ARCH models are ARMA-type processes for conditional heteroskedasticity, it is also natural to approach seasonality not only through seasonal lags as in (1.20) but also through seasonal variation in the parameters of regular ARCH models. This approach leads to periodic GARCH processes. A representative member of this class of processes can be written as

$$\varepsilon_t^2 = \mu + \phi_s \varepsilon_{t-1}^2 + v_t + \theta_s v_{t-1}, \tag{1.21}$$

where $y_t = \varepsilon_t$, v_t is i.i.d. $N(0, 1)$ while ϕ_s and θ_s are periodically varying coefficients. In the lower panels of Figure 1.10, a simulation example is plotted with a two-period cycle $a_1 = 0.5$ and $\alpha_2 = 0.9$ for $s = 1, 2$. A large sample of 250 and one of 1000 observations appear in Figure 1.10. We chose a cycle of two periods for illustrative purposes. To be more realistic, we would have to choose periodicity of say $s = 5$ for daily data. Periodic ARCH models have many properties of regular ARCH models and obviously also share features proper to periodic ARMA processes.

2 Deterministic Seasonality

2.1 Introduction

Chapter 1 has given an overview of various types of seasonal processes. This chapter looks at deterministic seasonality in more detail, commencing in Section 2.2 with the dummy variable and trigonometric representations. It is, however, impossible to consider recent developments concerned with deterministic seasonality in isolation from issues surrounding stochastic, and particularly nonstationary stochastic, seasonality. Therefore, our discussion continues in Section 2.3 with a further analysis and comparison between these processes introduced in Chapter 1. The final part of this chapter, Section 2.4, discusses recently proposed tests designed to distinguish between deterministic and nonstationary stochastic seasonality by testing the null hypothesis that seasonality is of the deterministic type. The issue of deterministic versus nonstationary stochastic seasonality reemerges in Chapter 3, where the nonstationary null hypothesis is examined. Thus, the analysis of much of the present chapter and that of Chapter 3 can be seen as complementary.

Some notation will assist in our discussion. When issues related to seasonality are considered, it is often convenient to recognize explicitly the season and the year to which a specific observation t relates. This is conveniently achieved by using two subscripts for a variable as in Section 1.3, with the first subscript referring to the season and the second to the year. With the simplifying assumption made that the first observation for $t = 1$ relates to season $s = 1$, then the double subscript notation writes the sequence of observations $y_1, y_2, \ldots, y_S, y_{S+1}, \ldots, y_{2S}, y_{2S+1}, \ldots$ as $y_{11}, y_{21}, \ldots, y_{S1}, y_{12}, \ldots, y_{S2}, y_{13}, \ldots$. In general, y_t is equivalently written as $y_{s\tau}$, where $s = 1 + [(t-1) \bmod S]$ and $\tau = 1 + \text{int}[(t-1)/S]$. With T observations available on y_t, then (again for simplicity of notation) we will assume that there are precisely T_τ years of data, so that $T_\tau = T/S$. It should be clear from the context whether the conventional single subscript notation y_t or the double subscript notation $y_{s\tau}$ is being used.

19

2.2   Representations of Deterministic Seasonality

Deterministic seasonality describes behavior in which the unconditional mean of the process varies with the season of the year. In this section we revisit and detail further the two representations of deterministic seasonality outlined in Chapter 1.

2.2.1   The Dummy Variable Representation

The conventional dummy variable representation of seasonality, as very frequently used in applied econometric studies, can be written as

$$y_t = \sum_{s=1}^{S} \gamma_s \delta_{st} + z_t, \qquad t = 1, \ldots, T_\tau, \tag{2.1}$$

where z_t is a weakly stationary stochastic process with mean zero and, as in Chapter 1, δ_{st} $(s = 1, \ldots, S)$ are seasonal dummy variables. Thus, for season s of year τ,

$$E(y_{s\tau}) = \gamma_s, \qquad s = 1, \ldots, S. \tag{2.2}$$

The property of (2.2) is the one of primary interest for deterministic seasonality, as it implies that the process has a seasonally shifting mean. Because of this time-varying mean, the process y_t is not stationary. This nonstationarity is often overlooked because it is simple to remove by subtraction of the mean for each season. Thus, the deviations $y_t - E(y_t) = z_t$ are weakly stationary, so that the second-order properties of these deviations are invariant to the season s in addition to the year τ.

One disadvantage of the simple dummy variable representation of (2.1), however, is that it mixes seasonality and the overall mean when the latter is nonzero. Since a proportion $1/S$ of observations relates to each of the S seasons, then the overall mean of y_t is

$$E(y_t) = \mu = \frac{1}{S} \sum_{s=1}^{S} \gamma_s. \tag{2.3}$$

When this overall mean is extracted, the deterministic seasonal effect for season s is $m_s = \gamma_s - \mu$. Obviously, this definition imposes the restriction $\sum_{s=1}^{S} m_s = 0$, with the natural interpretation that there is no deterministic seasonality when observations are summed over a year. When the level of the series (represented here by μ) is separated from the seasonal component, then

$$y_t = \mu + \sum_{s=1}^{S} m_s \delta_{st} + z_t. \tag{2.4}$$

It might be noted that the discussion of deterministic seasonality in Chapter 1 implicitly assumed $\mu = 0$.

When μ is replaced by $\mu_0 + \mu_1 t$, the last equation can be generalized to include a trend component that is constant over the seasons. A further generalization would allow a separate trend for each season, so that

$$y_t = \mu_0 + \mu_1 t + \sum_{s=1}^{S} (m_{0s} + m_{1s} t)\delta_{st} + z_t. \tag{2.5}$$

Once again, the seasonal coefficients are restricted so that $\sum_{s=1}^{S} m_{0s} = \sum_{s=1}^{S} m_{1s} = 0$ to ensure that seasonality sums to zero over the year. Such trending deterministic seasonality may, however, be unrealistic in practice for many economic series, since it implies that observations for the seasons of the year diverge over time. Therefore, any trend included is typically taken to be non-seasonal.

Whether in the process (2.4) or (2.5), one implication of stationarity for $z_t = y_t - E(y_t) = y_{s\tau} - E(y_{s\tau})$ is that each observation deviates from its respective seasonal mean with a variance that is constant over both s and τ. In this sense, the deterministic seasonal process ensures that the observations cannot "wander" too far from their underlying mean.

2.2.2 The Trigonometric Representation

Although perhaps less intuitive, it is sometimes useful to adopt the trigonometric representation of deterministic seasonality. As noted in Chapter 1, the deterministic seasonal process can equivalently be written in terms of trigonometric functions. Corresponding to (2.4), we have

$$y_t = \mu + \sum_{k=1}^{S/2} \left[\alpha_k \cos\left(\frac{2\pi kt}{S}\right) + \beta_k \sin\left(\frac{2\pi kt}{S}\right) \right] + z_t, \tag{2.6}$$

which also explicitly recognizes the overall mean μ. Notice that we need to consider α_k for $k = 1, \ldots, S/2$ but β_k only for $k = 1, \ldots, (S/2) - 1$, since $\beta_{S/2}$ multiplies a sine term that is always zero. This representation is the basis of spectral analysis of seasonality and seasonal adjustment; an early reference to this is Hannon, Terrell, and Tuckwell (1970).

For the case of quarterly data, (2.6) implies that the seasonal dummy variable coefficients of (2.1) are related to the deterministic components of the trigonometric representation by

$$\begin{aligned}
\gamma_1 &= \mu + \beta_1 - \alpha_2, \\
\gamma_2 &= \mu - \alpha_1 + \alpha_2, \\
\gamma_3 &= \mu - \beta_1 - \alpha_2, \\
\gamma_4 &= \mu + \alpha_1 + \alpha_2.
\end{aligned} \tag{2.7}$$

Notice that, in this quarterly case, α_1 and β_1 each give rise to a component of y_t that has a half-cycle every two periods and a full cycle every four periods. That is, these components cycle each year. The coefficients α_1 and β_1 are associated with the spectral frequency $\pi/2$, since these multiply $\cos(t\pi/2)$ and $\sin(t\pi/2)$ respectively, for $t = 1, 2, \ldots$ (note that it is a property of these trigonometric functions that they cycle every four periods through the values $1, 0, -1, 0$). Further, it might be remarked that although α_1 is associated with the second and fourth quarters in (2.7), while β_1 is associated with the first and third quarters, these two cycles cannot be distinguished in the frequency domain because they both correspond to four-period cycles.

Continuing the logic, the coefficient, α_2 is associated with the spectral frequency π, since it is the coefficient of terms of the form $\cos(t\pi)$ for $t = 1, 2, \ldots$ in (2.6). As these terms alternate between -1 and 1, α_2 contributes a full cycle every two periods. With quarterly data, the spectral frequencies $\pi/2$ and π are referred to as the seasonal frequencies, because any deterministic seasonal pattern over the four quarters of the year can be expressed as a linear function of terms at these two frequencies, namely as $\alpha_1 \cos(t\pi/2) + \beta_1 \sin(t\pi/2) + \alpha_2 \cos(t\pi)$. By construction (through these trigonometric functions), the seasonal pattern necessarily sums to zero over any four consecutive values of t.

Note that (2.7) can be written more compactly as

$$\Gamma = RB, \tag{2.8}$$

where $\Gamma = (\gamma_1, \gamma_2, \gamma_3, \gamma_4)'$, $B = (\mu, \alpha_1, \beta_1, \alpha_2)'$, and

$$R = \begin{bmatrix} 1 & 0 & 1 & -1 \\ 1 & -1 & 0 & 1 \\ 1 & 0 & -1 & -1 \\ 1 & 1 & 0 & 1 \end{bmatrix}. \tag{2.9}$$

This 4×4 nonsingular matrix R captures the one-to-one relationship between the trigonometric representation (2.6) and the more usual dummy variable representation in (2.1) for the quarterly case. The general matrix equation, (2.8), also, of course, applies for data sampled at other frequencies. In particular, the elements of R and B for monthly data can be defined from (2.6) with $S = 12$. In that case the seasonal frequencies are $\pi/6$, $\pi/3$, $\pi/2$, $2\pi/3$, $5\pi/6$, and π, with two cycles occurring at each of these frequencies (one through each of the cosine and the sine terms) except for the frequency π, where only the term $\cos(t\pi)$ applies. Any deterministic seasonal pattern in monthly data can be written by using the trigonometric cosine and sine functions at these seasonal spectral frequencies. Note that, once again by construction, the representation separates the overall mean μ from the deterministic seasonal component, with the latter necessarily summing to zero over any twelve consecutive values of t.

Returning to the general case of S seasons, the matrix R in (2.8) has some useful properties. With the overall mean included in the vector B, this is a square matrix and, since there is a one-to-one relationship between the seasonal dummy variable representation considered in Subsection 2.2.1 and the trigonometric representation, this matrix must be nonsingular. Indeed, the columns of this matrix are orthogonal to each other. That is, when the ith column is denoted as the vector R_i, so that $R = (R_1, \ldots, R_S)$, then $R_i' R_j = 0, i \neq j$. This property ensures that $R'R = D$ is a diagonal matrix. Thus, if the ith diagonal element of D is d_i, then

$$
R^{-1} = \begin{bmatrix} 1/d_1 R_1' \\ 1/d_2 R_2' \\ \vdots \\ 1/d_S R_S' \end{bmatrix}, \tag{2.10}
$$

so that the inverse of R (after appropriate scaling row by row) is the transpose of itself. For example, in the quarterly case it can be verified that

$$
R^{-1} = \begin{bmatrix} 0.25 & 0.25 & 0.25 & 0.25 \\ 0 & -0.5 & 0 & 0.5 \\ 0.5 & 0 & -0.5 & 0 \\ -0.25 & 0.25 & -0.25 & 0.25 \end{bmatrix}.
$$

Note also that the first column of R is a vector of ones. In this case, $R_1' R_1 = d_1 = S$ so that each element of the first row of R^{-1} is $1/S$. Consequently, the inverse yields the definitional relationship $\mu = (1/S) \sum_{s=1}^{S} \gamma_s$. It is also worth noting that (1.3)–(1.5) of Chapter 1, which express the coefficients α_k and β_k as cosine and sine functions, effectively present the elements of R^{-1}.

It is sometimes useful to associate the overall mean with the zero spectral frequency. This association follows as μ can be written in terms of trigonometric functions as $\alpha_0 \cos(2\pi kt/S)$ with $k = 0$, and hence (2.6) is equivalent to

$$
y_t = \sum_{k=0}^{S/2} \left[\alpha_k \cos\left(\frac{2\pi kt}{S}\right) + \beta_k \sin\left(\frac{2\pi kt}{S}\right) \right] + z_t, \tag{2.11}
$$

since $\sin(0) = 0$. However, having recognized that the overall mean has a spectral interpretation, we will continue to write it as μ and hence typically use the representation (2.6) in preference to (2.11).

Although we will not consider it explicitly, it should be clear that the coefficients of the trend in the seasonally varying trend model of (2.5) can also be written by using a trigonometric representation. With appropriate definition of the elements of Γ, R, and B in (2.8), this equation applies also to the relationships between the trend coefficients in the dummy variable and trigonometric representations.

2.3   Stochastic and Deterministic Seasonality

With deterministic seasonality now spelled out in some detail, we begin to explore some of the relationships between the various seasonal mean processes outlined in Chapter 1.

2.3.1   Stochastic Seasonal Processes

The discussion of Section 2.2 concentrated on deterministic seasonality that is evident through the unconditional mean, $E(y_{s\tau})$. Now we turn our attention to stochastic seasonality.

Specifically, consider the seasonal AR(1) process for z_t that was introduced in Chapter 1. When written in our double subscript notation, this becomes

$$z_{s\tau} = \phi_S z_{s,\tau-1} + \varepsilon_{s\tau}, \qquad s = 1, \ldots, S, \quad \tau = 1, 2, \ldots, T_\tau,$$

(2.12)

where $\varepsilon_t = \varepsilon_{s\tau}$ is i.i.d. $(0, \sigma^2)$. This form is useful in emphasizing that the autoregressive relationship for $z_{s\tau}$ in season s relates to the same season in the preceding year. It has already been noted in Chapter 1 that, conditional on past z, the mean exhibits seasonal patterns. Taking this further, by substituting for lagged z on the right-hand side of (2.12), we have

$$z_{s\tau} = \phi_S^\tau z_{s0} + \sum_{j=0}^{\tau-1} \phi_S^j \varepsilon_{s,\tau-j}.$$

(2.13)

Thus, this stochastic seasonal process contains two sources of seasonality. First, there is the influence of the unobserved starting value for season s, namely z_{s0}, which affects the subsequent observations for that season through $\phi_S^\tau z_{s0}$. Second, disturbances for the specific season s in previous years ($\varepsilon_{s,\tau-j}$ for $j > 0$) influence $z_{s\tau}$, so that patterns that occur by chance through the disturbances tend to be repeated. In both cases, the repeating patterns usually associated with seasonality require $\phi_S > 0$. When the process is also stationary, so that $0 < \phi_S < 1$, the effects of z_{s0} and of any specific $\varepsilon_{s,\tau-j}$ diminish over time.

The variance of (2.12) is

$$\mathrm{Var}\,(z_{s\tau}) = \phi_S^{2\tau} \mathrm{Var}(z_{s0}) + \sigma^2 \sum_{j=0}^{\tau-1} \phi_S^{2j}.$$

If the process is stationary, so that $-1 < \phi_S < +1$ and $\text{Var}(z_{s\tau}) = \text{Var}(z_{s0})$, then the variance is equal to

$$\text{Var}(z_{s\tau}) = \sigma^2 / (1 - \phi_S^2), \tag{2.14}$$

which is constant over both seasons $s = 1, \ldots, S$ and years $\tau = 1, 2, \ldots, T_\tau$.

Although stochastic seasonality is traditionally associated with processes containing annual lag(s), such as (2.12), seasonality can also arise when the stochastic process is a stationary autoregressive process of order less than S. For example, consider the AR(1) process

$$z_t = \phi z_{t-1} + \varepsilon_t, \qquad t = 1, 2, \ldots, T, \tag{2.15}$$

which implies

$$z_t = \phi^t z_0 + \sum_{j=0}^{t-1} \phi^j \varepsilon_{t-j}. \tag{2.16}$$

This process is nonseasonal if $0 < \phi < 1$, because the effects of z_0 and of any past disturbance are monotonically decreasing as t increases irrespective of the season. However, when $-1 < \phi < 0$, then z_t has a transient two-period pattern. For positive values of t, $\phi^t z_0$ is of the same sign as z_0, while for negative t, $\phi^t z_0$ is of the opposite sign (with the magnitude decreasing as t increases in both cases). In a similar way, the effect of any specific disturbance ε has a (declining) two-period pattern over time. If the data are quarterly, then such a half-year pattern can be considered as a particular form of stochastic seasonality. Another form of a seasonal pattern for quarterly data would arise through the AR(2) process

$$z_t = \phi_2 z_{t-2} + \varepsilon_t, \tag{2.17}$$

with $-1 < \phi_2 < 0$, where the pattern tends to alternate each two periods and repeat each four periods (that is, each year).

One further insight might help. The polynomial operator for the seasonal AR(1) process of (2.12) can be factorized in the case of quarterly data as

$$(1 - \phi_4 L^4) = (1 - \sqrt[4]{\phi_4}L)(1 + \sqrt[4]{\phi_4}L)(1 + \sqrt[2]{\phi_4}L^2), \tag{2.18}$$

when ϕ_4 is positive. Thus, the stationary quarterly seasonal AR(1) polynomial with positive coefficient can be decomposed into a nonseasonal factor $(1 - \sqrt[4]{\phi_4}L)$, together with two factors which contribute to seasonality, namely $(1 + \sqrt[4]{\phi_4}L)$ and $(1 + \sqrt[2]{\phi_4}L^2)$. These latter two factors play roles analogous to those of the alternating pattern each period (captured through α_2) and the two semiannual patterns (captured through α_1 and β_1) in the quarterly deterministic seasonality case of (2.7). Indeed, the autoregressive processes $(1 + \sqrt[4]{\phi_4}L)y_t = \varepsilon_t$ and $(1 + \sqrt[2]{\phi_4}L^2)y_t = \varepsilon_t$ have peaks in their spectral densities at frequencies π and $\pi/2$, respectively. It might also be noted that

$(1 + \sqrt[2]{\phi_4}L^2)$ can be factorized as $(1 + i\sqrt[4]{\phi_4}L)(1 - i\sqrt[4]{\phi_4}L)$ where $i = \sqrt{-1}$. This complex pair of factors associated with the frequency $\pi/2$ cannot be separated, since the factor must occur together in order to give rise to a real-valued process.

There is a corresponding factorization to (2.18) for the monthly process $(1 - \phi_{12}L^{12})$ with ϕ_{12} positive. In that case, the factors of the form $(1 + \psi_i L)$ $(i = 1, \ldots, 12)$ comprise a nonseasonal factor and eleven seasonal factors. The seasonal factors form five complex pairs, which are associated with the seasonal frequencies $\pi/6, \pi/3, \pi/2, 2\pi/3$, and $5\pi/6$, together with a real factor associated with the frequency π. These are, of course, the seasonal frequencies also noted for the monthly case in the discussion of deterministic seasonality (see Subsection 2.2.2).

2.3.2 The Seasonal Random Walk

Most of the discussion of the previous subsection applies also to the case of a seasonal random walk, where $\phi_S = 1$. Here (2.12) and (2.13), applied to the observed process $y_{s\tau}$, then become

$$y_{s\tau} = y_{s,\tau-1} + \varepsilon_{s\tau},$$

$$= y_{s0} + \sum_{j=1}^{\tau} \varepsilon_{sj}. \qquad (2.19)$$

Although a detailed analysis of this process will be left to the next chapter, some preliminary examination is relevant for the discussion here. Note, first, that (2.19) applies for $s = 1, \ldots, S$. In other words, the seasonal random walk process consists of S random walks, one associated with each season of the year. Since the disturbances are independent over the seasons, these S random walks are independent of each other.

In contrast with the stationary stochastic seasonal process, for the seasonal random walk, the influence of the starting value for season s (namely y_{s0}) and any specific disturbance for season s (ε_{sj}) does not diminish as the year τ increases. Therefore, if $E(y_{s0}) = \gamma_s \neq 0$, then $E(y_{s\tau}) = \gamma_s$ for all τ. In other words, any deterministic seasonal component in the starting value for the seasonal random walk in season s is carried over to all subsequent observations relating to that season.[1] This property holds also for (nontrending) deterministic seasonality. Nevertheless, there is an important distinction between the deterministic seasonal process of Section 2.2 and the seasonal random walk process

---

[1] It is sometimes erroneously said that the seasonal random walk has no deterministic component. This confusion arises because the assumption $y_{s0} = 0$, or at least $E(y_{s\tau}) = 0$, is frequently made for simplicity.

examined here. That is, in the former case, Var $(y_{s\tau})$ is constant over both s and τ. In (2.19), however,

$$\text{Var}(y_{s\tau}) = \text{Var}(y_{s0}) + \tau\sigma^2,$$

which linearly increases with the year τ. As a result of this increasing variance, the observation $y_{s\tau}$ can wander far from its unconditional mean γ_s over time.

For the quarterly seasonal random walk process, the factorization (2.18) of the autoregressive polynomial becomes

$$(1 - L^4) = (1 - L)(1 + L)(1 + L^2). \qquad (2.20)$$

This factorization, in turn, implies that the spectral density of y_t,

$$f_y(\omega) = \frac{\sigma^2}{|1 - e^{-4i\omega}|^2} = \frac{\sigma^2}{|(1 - e^{-i\omega})(1 + e^{-i\omega})(1 + e^{-2i\omega})|^2},$$

has infinite power at the zero frequency (because of the conventional unit root of 1) and at the seasonal frequencies of $\omega = \pi$ and $\omega = \pi/2$ (corresponding to the unit roots of -1 and $\pm i$ arising from the factors $1 + L$ and $1 + L^2$, respectively). Similarly, the corresponding factorization of $(1 - L^{12})$ for the monthly case gives rise to infinite spectral power at the zero frequency plus the monthly seasonal frequencies.

The seasonal random walk will play a fundamental role in the discussions of the next chapter. At this preliminary stage, we assume that the process has starting values $y_{s0} = 0$ ($s = 1, \ldots, S$). Then, as $T_\tau \to \infty$, the behavior of each scaled partial sum $T_\tau^{-1/2} y_{s\tau} = T_\tau^{-1/2} \sum_{j=1}^{\tau} \varepsilon_{sj}$ ($s = 1, \ldots, S$) converges to Brownian motion. That is,

$$T_\tau^{-1/2} y_{s\tau} \Rightarrow \sigma W_s(r), \qquad s = 1, \ldots, S, \qquad (2.21)$$

where $W_s(r)$ is standard Brownian motion, derived from an i.i.d. $(0, 1)$ disturbance, and $\Rightarrow$ indicates convergence in distribution. Using well-known results for Brownian motion [see, e.g., Hamilton (1994a) or Banerjee, Lumsdaine, and Stock (1992)], we can also note that

$$T_\tau^{-3/2} \sum_{\tau=1}^{T_\tau} y_{s\tau} \Rightarrow \sigma \int_0^1 W_s(r)\, dr, \qquad s = 1, \ldots, S, \qquad (2.22)$$

$$T_\tau^{-1} \sum_{\tau=1}^{T_\tau} \varepsilon_{s\tau} y_{s,\tau-1} \Rightarrow \sigma^2 \int_0^1 W_s(r)\, dW_s(r), \qquad s = 1, \ldots, S, \qquad (2.23)$$

$$T_\tau^{-2} \sum_{\tau=1}^{T_\tau} y_{s\tau}^2 \Rightarrow \sigma^2 \int_0^1 [W_s(r)]^2\, dr, \qquad s = 1, \ldots, S. \qquad (2.24)$$

Notice that the Brownian motions in (2.21)–(2.24) are indexed by the season s and, since the disturbances $\varepsilon_{s\tau}$ underlying these are independent over seasons, $W_s(r)$ are also independently distributed over $s = 1, \ldots, S$.

The seasonal random walk process of (2.19) can be extended to a more general seasonal unit root process. In this case, $\Delta_S y_t = z_t$ is a stationary and invertible ARMA process. Discussion of this case is, however, left to Chapter 3.

2.3.3 Deterministic Seasonality versus Seasonal Unit Roots

The discussion so far has hinted at an issue that is central to much of the recent literature on seasonality, namely whether an observed series should be modeled as a deterministic seasonal process or as a seasonal unit root process, in either case possibly also with a stationary stochastic seasonal component. For finite T, the two processes can have similar empirical properties, with both having strong seasonal patterns and empirical spectral power concentrated at the seasonal frequencies. In this chapter, we consider tests of the deterministic seasonality null hypothesis (Section 2.4). Tests of the seasonal unit root null hypothesis are discussed in Chapter 3. Before we move on to consider the deterministic seasonality tests, however, it is important to appreciate the relationship between the competing hypotheses.

Bell (1987) considers this relationship in the context of monthly time series, but his analysis can be readily applied to any S. He considers two competing specifications, one the simple deterministic seasonality model,

$$y_{s\tau} = \gamma_s + \varepsilon_{s\tau}, \qquad s = 1, \ldots, S, \qquad \tau = 1, \ldots, T_\tau, \qquad (2.25)$$

and the other the process

$$\Delta_S y_{s\tau} = (1 - \theta_S L^S)\varepsilon_{s\tau}, \qquad s = 1, \ldots, S, \qquad \tau = 1, \ldots, T_\tau. \quad (2.26)$$

Note that for $-1 < \theta_S < 1$, (2.26) defines a specific seasonal unit root process that is a seasonal MA process in $\Delta_S y_t$. More importantly, Bell shows that these two processes are equivalent in the special case where $\theta_S = 1$.

The basic result is easy to obtain from a straightforward extension to our analysis above. When $\varepsilon_{s\tau}$ is replaced by $(1 - \theta_S L^S)\varepsilon_{s\tau}$ in (2.19), it is simple to see that the annual differenced process (2.26) implies

$$
\begin{aligned}
y_{s\tau} &= y_{s,\tau-1} + (1 - \theta_S L^S)\varepsilon_{s\tau}, \\
&= y_{s0} + \sum_{j=1}^{\tau}(1 - \theta_S L^S)\varepsilon_{sj}.
\end{aligned}
\qquad (2.27)
$$

For $\theta_S = 1$, this becomes

$$
\begin{aligned}
y_{s\tau} &= y_{s0} + \sum_{j=1}^{\tau}(\varepsilon_{sj} - \varepsilon_{s,j-1}), \\
&= y_{s0} - \varepsilon_{s0} + \varepsilon_{s\tau}, \\
&= \gamma_s + \varepsilon_{s\tau},
\end{aligned}
$$

when $y_{s0} = \gamma_s + \varepsilon_{s0}$. This last line is, of course, the deterministic seasonality equation (2.25). Thus, with the identical assumption $y_{s0} = \gamma_s + \varepsilon_{s0}$ made about the starting values, the two processes of (2.25) and (2.26) with $\theta_S = 1$ lead to precisely the same expression for any $y_{s\tau}$, and hence are equivalent. This simple result also applies when $\varepsilon_{s\tau}$ is replaced by the more general stationary ARMA process $z_{s\tau}$.

It is often said that the seasonal differencing operator Δ_S and the noninvertible MA operator $1 - L^S$ cancel in (2.26) with $\theta_S = 1$. The logic may be extended to show that when "near cancellation" occurs, with θ_S close to but less than unity in (2.26), then the properties of this seasonal unit root process for finite T are empirically similar to those of the deterministic seasonal process of (2.25). Conversely, when the latter process includes a stationary seasonal AR term with a coefficient close to but less than unity, its properties may be empirically like those of a seasonal MA process in the seasonal difference $\Delta_S y_t$. In other words, for the sample sizes often observed in practice, it may be difficult to discriminate between deterministic seasonality and a seasonal unit root process.

2.3.4 Unobserved Components Approaches

In the so-called basic structural model of Harvey (1989), the seasonal component is rationalized through the deterministic seasonal process of (2.1). In particular, instead of seasonality summing to zero over a year in a deterministic way, the seasonal component summed over a year is specified to be random with a zero mean. Thus, as in equation (1.16) of Chapter 1,

$$
(1 + L + \cdots + L^{S-1})y_t^s = \varepsilon_t^s, \tag{2.28}
$$

with ε_t^s being i.i.d. $(0, \omega^2)$ and independent of the disturbances driving the other components. The deterministic seasonal model then becomes a special case of (2.28) with $\omega^2 = 0$. By the addition of a disturbance term with nonzero variance, this unobserved components approach allows seasonality to evolve over time.

However, the addition of this disturbance term is not innocuous in its implications, since the seasonal component then becomes a nonstationary process. Indeed, the autoregressive process for y_t^s in (2.28) has $S - 1$ unit roots, with these roots occurring at the seasonal frequencies.

A different form for a nonstationary stochastic seasonal component arises if the coefficients α and β of the trigonometric representation of Subsection 2.2.2 are allowed to evolve as random walks. In this case,

$$
\begin{aligned}
\alpha_{kt} &= \alpha_{k,t-1} + \eta_{kt}, \qquad k = 1, \ldots, S/2, \\
\beta_{kt} &= \beta_{k,t-1} + \eta_{kt}^*, \qquad k = 1, \ldots, (S/2) - 1,
\end{aligned}
\tag{2.29}
$$

where all η_{kt} and η_{kt}^* are i.i.d. $(0, \omega^2)$ processes. This generalization of deterministic seasonality is adopted by Canova and Hansen (1995), and views the seasonality as evolving over time through evolution of the components at each of the seasonal frequencies $2\pi k / S$ $(k = 1, \ldots, S/2)$. The α_k and β_k are definitionally associated with the seasonal frequencies, so that the nonstationarity arising from (2.29) for a given k is directly associated with the corresponding seasonal frequency.

As just discussed, the unobserved component approach often generalizes deterministic seasonality to allow it to evolve over time. Such evolution, however, also generally implies that seasonality is a nonstationary stochastic process with unit roots at all seasonal frequencies. This is not to say that it is impossible to define an unobserved components model with stationary seasonality; such models are discussed (for example) at some length by Nerlove et al. (1995). Rather, the unobserved components models of interest in recent research, including research related to seasonal adjustment as discussed in Chapter 4, are based on nonstationary seasonality.

Whether it is in combination with (2.28) or (2.29), the unobserved components approach also typically specifies a nonstationary stochastic process for the nonseasonal component. The simplest case of such a process is, of course, the random walk, whereby the nonseasonal component is $y_t^{ns} = \mu_t$ with

$$
\mu_t = \mu_{t-1} + \varepsilon_t^{ns}.
\tag{2.30}
$$

2.4   Testing Deterministic Seasonality

It is obvious from the discussion of Section 2.3 that, although a deterministic seasonal representation is frequently used in applied econometric practice, there is an important issue of how one might validly test whether seasonality is, indeed, of this type.

In the context of the unobserved components model, Harvey (1989) proposes testing the null hypothesis of deterministic seasonality by means of a test of $\omega^2 = 0$ against $\omega^2 > 0$ in the context of either (2.28) or (2.29). This test can be conducted by using conventional likelihood ratio, Wald or Lagrange multiplier test principles, provided that account is taken of the fact that the parameter is on the boundary (zero) of the permissible parameter space under the null hypothesis. While this approach is attractive in its simplicity, it is also restrictive because it is based on the assumption that the specified (relatively simple) unobserved components model adequately represents the data-generating process for y_t. In particular, no stationary stochastic seasonality is allowed. While still testing within the broad framework of an unobserved components approach, the recent test of Canova and Hansen allows more general types of processes.

2.4.1 The Canova–Hansen Test

The approach of Canova and Hansen (1995) is based on the trigonometric representation of deterministic seasonality, as given by (2.6). This representation has been discussed at some length in Subsection 2.2.2 and, for the purposes here, we rewrite it as

$$y_t = \sum_{s=1}^{S} F_s' B_t \delta_{st} + z_t, \tag{2.31}$$

where the $1 \times S$ vector F_s' is the sth row of the matrix R defined for the quarterly case in (2.9) and $B_t = (\mu_t, \alpha_{1t}, \beta_{1t}, \ldots, \alpha_{S/2,t})'$. The disturbances z_t are assumed to be normally distributed and stationary, but not necessarily uncorrelated over time. When $B_t = B$ for all t, then this equation is identical to (2.6). However, the test is based on allowing the possibility that all elements of B evolve according to a (vector) random walk, such that

$$B_t = B_{t-1} + V_t,$$

$$= B_0 + \sum_{i=1}^{t} V_i. \tag{2.32}$$

It is assumed that V_t is i.i.d. with $E(V_t V_t') = \omega^2 H$, where H is a known positive definite matrix and the vector disturbance V_t is independent of z_t. Under the null hypothesis, $\omega^2 = 0$, so that B is constant over time and hence deterministic seasonality is present. The alternative hypothesis is $\omega^2 > 0$, which implies that each element of B_t contains a unit root. Thus, because the elements of B are associated with the zero and seasonal frequencies, the alternative hypothesis is nonstationarity for y_t at all these frequencies. In the terminology of Chapter 3, the process is seasonally integrated under this alternative.

It might be noted at this stage that Canova and Hansen discuss only the parameters α_k and β_k associated with seasonality evolving over time, but, as recognized by Hylleberg (1995), the analysis can be extended to include the constant μ. We include the constant in our definition of the vector B, but we later return to consideration of the seasonal parameters alone.

Under the Canova and Hansen null hypothesis of deterministic seasonality, the data-generating process, or DGP, of (2.31) with (2.32) can be written in vector notation as

$$Y_\tau = \Gamma + Z_\tau = RB + Z_\tau, \tag{2.33}$$

where $Y_\tau = (y_{1\tau}, \ldots, y_{S\tau})'$ is the vector of observations for year τ and Z_τ is the corresponding zero mean stationary disturbance process whose covariance matrix can be represented as $\Omega_Z = E(Z_\tau Z_\tau')$. In addition to autocorrelation, Ω_Z might allow heteroskedasticity over seasons. The matrix R is identical to that discussed above and, as noted in Subsection 2.2.2, the columns of this matrix are mutually orthogonal so that $R'R = D$ is a diagonal matrix. Therefore, premultiplying (2.33) by R' yields

$$R'Y_\tau = DB + R'Z_\tau, \tag{2.34}$$

with $E(R'Z_\tau Z_\tau' R) = \Omega_{RZ} = R'\Omega_Z R$. Since D is diagonal, DB is a scaled version of B. In the quarterly case, for example, $DB = (4\mu, 2\alpha_1, 2\beta_1, 4\alpha_2)'$. The form (2.34) implies that the ith element of $R'Y_\tau$ is uniquely associated with the ith element of B (multiplied by d_i), with disturbance given by the ith element of $R'Z_\tau$. For technical reasons (in order to obtain a test statistic free of nuisance parameters), the covariance matrix H of V_t in (2.32) and used in the Canova–Hansen test is assumed to be related to the covariance matrix of $R'Z_\tau$ such that $H = \Omega_{RZ}^{-1}$.

The dummy variable and trigonometric representations of deterministic seasonality are equivalent. Therefore, ordinary least-squares (OLS) residuals for (2.1) and (2.31) are identical. These OLS residuals, which are obtained under the null hypothesis $\omega^2 = 0$, are used in construction of the test statistic. From the OLS residuals, $\widehat{z}_t$ $(t = 1, \ldots, T)$ form the $S \times 1$ vectors of partial sums $\widehat{Z}_t^a$, where $\widehat{Z}_t^{a'} = (\sum_{j=1}^t \widehat{z}_j \delta_{1j}, \ldots, \sum_{j=1}^t \widehat{z}_j \delta_{Sj})$ for $t = 1, \ldots, T$. Thus, the vector $\widehat{Z}_t^a$ aggregates over time periods to t the residuals for each of the seasons, with the sth element of $\widehat{Z}_t^a$ being the aggregated residuals for season s. Based on an analysis by Nyblom (1989) of locally most powerful tests for the null hypothesis of constant coefficients, Canova and Hansen suggest a Lagrange multiplier test statistic, which can be written for our case as

$$L = \frac{S}{T^2} \sum_{t=1}^T \left(R'\widehat{Z}_t^a \right)' \widehat{\Omega}_{RZ}^{-1} \left(R'\widehat{Z}_t^a \right),$$

$$= \frac{S}{T^2} \sum_{t=1}^T \widehat{Z}_t^{a'} \widehat{\Omega}_Z^{-1} \widehat{Z}_t^a, \tag{2.35}$$

where $\widehat{\Omega}_{RZ} = R'\widehat{\Omega}_Z R$ is computed by using an estimator that is consistent under the null hypothesis.

The form of the test statistic (2.35) is explained further below through a simple example for the quarterly case. To appreciate the asymptotic distribution of this statistic in the general case, note that for the annual disturbance vector Z_τ, the elements of $(\Omega_Z)^{-1/2} Z_\tau$ are mutually uncorrelated with common variance of unity (note, however, that they are not necessarily uncorrelated over time). With the assumption of normality, then it follows that the elements of $(\Omega_Z)^{-1/2} Z_\tau$ are also mutually independent. Aggregating over years 1 to τ yields $X_\tau = (\Omega_Z)^{-1/2} \sum_{j=1}^{\tau} Z_j$ as a vector of independent $I(1)$ processes, each driven by a disturbance with unit variance and where the starting value of each is zero. Then, analogously to (2.21),

$$\frac{1}{\sqrt{T_\tau}} X_\tau \Rightarrow W^x(r), \tag{2.36}$$

where $W^x(r)$ is an $S \times 1$ vector of independent standard Brownian motions. Further, analogously to (2.24),

$$\frac{1}{T_\tau^2} \sum_{j=1}^{T_\tau} X_\tau' X_\tau \Rightarrow \int_0^1 W^x(r)' W^x(r) \, dr \tag{2.37}$$

[see Hamilton (1994a) or the discussion of asymptotic distributions in Chapter 3].

Under the null hypothesis, the sample quantity of (2.35) converges to a distribution closely related to (2.37), namely

$$L \Rightarrow \int_0^1 [W^x(r) - r W^x(1)]' [W^x(r) - r W^x(1)] \, dr. \tag{2.38}$$

This last result can be viewed as involving a correction of the vector Brownian motion $W^x(r)$ by subtracting $r W^x(1)$. This correction arises because, in practice, B in (2.34) must be estimated; see Nyblom (1989). In addition, (2.35) also differs from the left-hand side of (2.37) in scaling by S/T^2 and in summation over all t. However, it should be recognized that $T = ST_\tau$ and also that summation over t implies ST_τ terms rather than T_τ. Hence, these latter changes have no effect on the asymptotic distribution.

The distribution of (2.38) is nonstandard and, for the $S \times 1$ vector $W^x(r)$, it is sometimes known as the von Mises distribution with S degrees of freedom, or $VM(S)$. This asymptotic distribution is tabulated by Nyblom (1989) and also by Canova and Hansen (1995).

In the Canova–Hansen test, the null hypothesis is rejected for large values of L. Under the alternative hypothesis, the residuals from (2.1) used to construct the test statistic reflect random walk behavior in B_t. The constructed partial sums in (2.35) then emphasize this behavior. Indeed, the partial sums in $R' \widehat{Z}_t^a$ are constructed so as to mimic (under the null hypothesis) the behavior of $\sum_{j=1}^t V_j$ under the alternative.

Analogous results to (2.38) apply when the overall mean μ is not included in the analysis, or when other subsets of B are considered. Essentially, such subsets of the vector B are considered by defining the random walk behavior of (2.32) to relate to the elements of interest and then constructing the test statistic

(2.35) corresponding to these. Clearly, the number of possible unit roots under test gives rise to the degrees of freedom for the von Mises distribution. Thus, for instance, if deterministic seasonality is tested excluding the overall mean, then $S - 1$ seasonal unit roots are allowed under the alternative hypothesis and the asymptotic distribution of L is $VM(S - 1)$. It should also be noted that (except for the seasonal term $\alpha_{S/2}$ corresponding to the frequency π), elements α_k and β_k of B must be considered together since these roots cannot be separately distinguished in their cyclical properties. If applied to μ alone, this effectively becomes the KPSS test of Kwiatkowski et al. (1992).

Canova and Hansen propose the use of the Newey and West (1987) nonparametric kernel estimator in order to obtain $\widehat{\Omega}_Z$. Their form of the test regression also allows explanatory variables. However, while they recommend that these explanatory variables should typically include y_{t-1}, Hylleberg (1995) points out that this should not be included, because y_{t-1} could absorb at least some of the impact of the seasonal unit root of -1. This seasonal unit root always arises when the DGP has the seasonally integrated form discussed at length in Chapter 3. When the zero frequency is included in the test statistic, then y_{t-1} could similarly absorb some of the zero frequency unit root of $+1$. Longer lags of the dependent variable also cannot be included, as these could absorb seasonal unit roots and hence render the test useless. Therefore, the Canova–Hansen approach is most appropriate where all serial correlation is handled nonparametrically and hence no lagged dependent variables are included in the test regression.

Example: The Simplest Quarterly Case

To clarify the above discussion, consider the case of quarterly data, for which R is given in (2.9). To keep the analysis as simple as possible, also assume that $z_t = \varepsilon_t$ is neither autocorrelated nor heteroskedastic, so that $E(Z_\tau Z_\tau') = \sigma^2 I_S$. For this special case, $\Omega_{RZ} = E(R' Z_\tau Z_\tau' R)$ is

$$\Omega_{RZ} = \sigma^2 R' R = \sigma^2 D,$$

$$= \sigma^2 \begin{bmatrix} 4 & 0 & 0 & 0 \\ 0 & 2 & 0 & 0 \\ 0 & 0 & 2 & 0 \\ 0 & 0 & 0 & 4 \end{bmatrix}. \tag{2.39}$$

Now, if it is known that the disturbance $z_t = \varepsilon_t$ has these standard properties under the null hypothesis, then OLS estimation of the deterministic seasonal component is optimal. To keep the analysis as simple as possible, suppose that this is exploited by undertaking OLS estimation in the dummy variable representation of (2.1). Let the ordinary least-squares residuals be $\widehat{\varepsilon}_t = \widehat{\varepsilon}_{st}$, where the latter form uses the notation for season s of year τ. Further, making the simplifying assumption that period t is the fourth quarter of year τ, then

$$
R'\widehat{Z}_t^a = R'\widehat{E}_t^a =
\begin{bmatrix}
1 & 1 & 1 & 1 \\
0 & -1 & 0 & 1 \\
1 & 0 & -1 & 0 \\
-1 & 1 & -1 & 1
\end{bmatrix}
\begin{bmatrix}
\sum_{j=1}^{\tau} \widehat{\varepsilon}_{1j} \\
\sum_{j=1}^{\tau} \widehat{\varepsilon}_{2j} \\
\sum_{j=1}^{\tau} \widehat{\varepsilon}_{3j} \\
\sum_{j=1}^{\tau} \widehat{\varepsilon}_{4j}
\end{bmatrix},
$$

$$
=
\begin{bmatrix}
\sum_{j=1}^{\tau} \widehat{\varepsilon}_{1j} + \sum_{j=1}^{\tau} \widehat{\varepsilon}_{2j} + \sum_{j=1}^{\tau} \widehat{\varepsilon}_{3j} + \sum_{j=1}^{\tau} \widehat{\varepsilon}_{4j} \\
\sum_{j=1}^{\tau} \widehat{\varepsilon}_{4j} - \sum_{j=1}^{\tau} \widehat{\varepsilon}_{2j} \\
\sum_{j=1}^{\tau} \widehat{\varepsilon}_{1j} - \sum_{j=1}^{\tau} \widehat{\varepsilon}_{3j} \\
\sum_{j=1}^{\tau} \widehat{\varepsilon}_{4j} - \sum_{j=1}^{\tau} \widehat{\varepsilon}_{3j} + \sum_{j=1}^{\tau} \widehat{\varepsilon}_{2j} - \sum_{j=1}^{\tau} \widehat{\varepsilon}_{1j}
\end{bmatrix}. \quad (2.40)
$$

More generally, for a t which corresponds to season $s < 4$ of year τ, then the summations in (2.40) will be up to year $\tau - 1$ for quarters $s + 1, \ldots, 4$ as these are subsequent to t.

With our simplifying assumptions, then with the use of (2.40) and (2.39), the Canova–Hansen test statistic as given by the first line of (2.35) can be seen to be

$$
L = \frac{1}{T^2 \widetilde{\sigma}^2} \sum_{t=1}^{T} \left[\left(\sum_{j=1}^{\tau} \widehat{\varepsilon}_{1j} + \sum_{j=1}^{\tau} \widehat{\varepsilon}_{2j} + \sum_{j=1}^{\tau} \widehat{\varepsilon}_{3j} + \sum_{j=1}^{\tau} \widehat{\varepsilon}_{4j} \right)^2 \right.
$$

$$
+ 2 \left(\sum_{j=1}^{\tau} \widehat{\varepsilon}_{4j} - \sum_{j=1}^{\tau} \widehat{\varepsilon}_{2j} \right)^2 + 2 \left(\sum_{j=1}^{\tau} \widehat{\varepsilon}_{1j} - \sum_{j=1}^{\tau} \widehat{\varepsilon}_{3j} \right)^2
$$

$$
+ \left. \left(\sum_{j=1}^{\tau} \widehat{\varepsilon}_{4j} - \sum_{j=1}^{\tau} \widehat{\varepsilon}_{3j} + \sum_{j=1}^{\tau} \widehat{\varepsilon}_{2j} - \sum_{j=1}^{\tau} \widehat{\varepsilon}_{1j} \right)^2 \right], \quad (2.41)
$$

or

$$
L = L_0 + L_{\pi/2} + L_{\pi}, \quad (2.42)
$$

where $\widehat{\Omega}_{RZ} = \widetilde{\sigma}^2 D$ and $\widetilde{\sigma}^2$ is the usual OLS estimator of σ^2. In the latter form of (2.42), L_0, $L_{\pi/2}$, and L_{π} provide test statistics relating to the 0, $\pi/2$, and π frequencies, respectively. These three components are given by the expressions in the first, second, and third lines of (2.41), respectively (in each case summed over $t = 1, \ldots, T$ and scaled by division by $T^2 \widetilde{\sigma}^2$). Their relationship to the spectral frequencies follows from (2.34) and the discussion of Subsection 2.2.2. Tedious but straightforward algebra reveals that this test statistic can also be written as

$$L = \frac{4}{T^2 \widehat{\sigma}^2} \sum_{t=1}^{T} \left[\left(\sum_{j=1}^{\tau} \widehat{\varepsilon}_{1j} \right)^2 + \left(\sum_{j=1}^{\tau} \widehat{\varepsilon}_{2j} \right)^2 + \left(\sum_{j=1}^{\tau} \widehat{\varepsilon}_{3j} \right)^2 \right.$$
$$\left. + \left(\sum_{j=1}^{\tau} \widehat{\varepsilon}_{4j} \right)^2 \right], \tag{2.43}$$

which expresses the statistic in terms of separate squared partial sums for each of the four quarters. Thus, (2.43) corresponds to the form given for the general Canova–Hansen test statistic in the second line of (2.35).

Under the overall null hypothesis, $L \sim VM(4)$. A test of the null hypothesis of deterministic seasonality, excluding consideration of the constant μ (at the zero frequency), would be distributed as $VM(3)$. Separate tests at the spectral frequencies of $\pi/2$ and π are $VM(2)$ and $VM(1)$, respectively, under the null hypothesis.

In the more general quarterly case in which the disturbances of (2.31) are not i.i.d. $(0, \sigma^2)$, then the simple summation of the separate test statistics as in (2.42) no longer applies. This is because the structure of Ω_Z comes into play and hence the test statistics at different frequencies are no longer mutually independent.

2.4.2 The Caner Test

Caner (1998) adopts the Canova–Hansen framework, but he exploits the advantages that accrue when the disturbances of (2.31) are independent over time. Thus, the Caner test is, in spirit, the same as the simple i.i.d. case just discussed. To be more specific, Caner proposes that the dynamic properties of the time series process should be handled by a parametric autoregressive augmentation, instead of by use of the nonparametric Newey–West correction adopted by Canova and Hansen. Therefore, the disturbances in the Caner test regression are assumed to be i.i.d.

Caner considers a test for the null hypothesis of deterministic seasonality, excluding the overall constant μ from consideration. Nevertheless, his approach can be generalized to include this latter term. With this generalization, the null hypothesis model is

$$\phi(L)y_t = \sum_{s=1}^{S} F_s' B_t \delta_{st} + \varepsilon_t, \tag{2.44}$$

with B_t constant over t. From our earlier discussion (especially Subsection 2.3.3), it follows that, with appropriate assumptions about the starting values, the null hypothesis process is equivalent to

$$\phi(L)\Delta_S y_t = \theta(L)\varepsilon_t, \tag{2.45}$$

with $\theta(L) = 1 - L^S$. The alternative hypothesis is that $\theta(L)\varepsilon_t$ in (2.45) is a general MA(S) process and not $(1 - L^S)\varepsilon_t$. To be more specific, the alternative hypothesis implies that B_t contains one or more unit roots and hence y_t is nonstationary. If this nonstationarity has the form that B_t contains S unit roots, then Δ_S removes this nonstationarity and $\theta(L)\varepsilon_t = \varepsilon_t$. However, the general MA(S) polynomial $\theta(L)$ also allows the possibility under the alternative that some but not all elements of B_t evolve over time. In practical terms, Caner proposes estimation of $\phi(L)$ from (2.45), which permits consistent estimation under both the null and alternative hypotheses. In particular, he proposes the use of maximum likelihood estimation rather than least squares.

Analogous to our special case of the Canova–Hansen test statistic in (2.41), the test statistic of Caner is

$$L = \frac{S}{\widetilde{\sigma}^2 T^2} \sum_{t=1}^{T} \left(R'\widehat{E}_t^a\right)' D^{-1}\left(R'\widehat{E}_t^a\right), \tag{2.46}$$

where $\widetilde{\sigma}^2$ is a consistent estimator of $\sigma^2 = \mathrm{Var}\,(\varepsilon_t)$ and $\widehat{E}_t^a$ is constructed from the residuals of (2.44). Under the null hypothesis, $L \Rightarrow VM(S)$. As in (2.42), this overall statistic can be decomposed as the sum of statistics that test the null hypothesis of constant elements of B_t for the zero frequency and for each of the seasonal frequencies. Thus, if deterministic seasonality is tested with the overall mean excluded, then in the quarterly case we have the asymptotic result that the test statistic $L_{\pi/2} + L_{\pi} \Rightarrow VM(3)$.

2.4.3 The Tam–Reinsel Test

Tam and Reinsel (1997) also examine the validity of the deterministic seasonality null hypothesis. By taking annual differences in the dummy variable representation (2.1), their test is a test of the null hypothesis $\theta_S = 1$ in

$$\Delta_S y_t = z_t - \theta_S z_{t-S}, \qquad t = 1, \dots, T. \tag{2.47}$$

Although we will return to the case of a general stationary stochastic disturbance process z_t, our discussion initially assumes that $z_t = \varepsilon_t \sim$ i.i.d. $(0, \sigma^2)$. From a simple comparison of Caner's representation (2.45) with (2.47), it is evident that (at least in spirit) the approach is similar to that of Caner (1998). The relationship between these two approaches is explored further in Subsection 2.4.3.

Tam and Reinsel take the perspective of a locally best invariant unbiased test, and they present two equivalent forms for their test statistic. The first approach is through (2.47). With the i.i.d. disturbance assumption, the equation of interest under the null hypothesis (i.e., with $\theta_S = 1$) can be written as

$$\Delta_S y_{s\tau} = \varepsilon_{s\tau} - \varepsilon_{s,\tau-1}, \tag{2.48}$$

which makes clear that only observations and disturbances for season s are involved. The vector of these differenced values relating to year τ is $\Delta_S Y_\tau = (\Delta_S y_{1\tau}, \ldots, \Delta_S y_{S\tau})'$, and, when the null hypothesis is true, the covariance properties of this vector are given by

$$
E(\Delta_S Y_{\tau_1} \Delta_S Y'_{\tau_2}) = \begin{cases} 2\sigma^2 I_S, & \tau_2 = \tau_1 \\ -\sigma^2 I_S, & \tau_2 = \tau_1 \pm 1 \\ 0, & \text{otherwise} \end{cases}.
$$

If the complete sample period vector, namely $\Delta_S Y = (\Delta_S Y'_1, \ldots, \Delta_S Y'_{T_\tau})'$, is considered, then its $T \times T$ covariance matrix is consequently known under the null hypothesis. When we denote $E(\Delta_S Y \Delta_S Y') = \Omega_Y$ for $\theta_S = 1$, the locally best invariant unbiased test statistic for the null hypothesis $\theta_S = 1$ against the invertible seasonal moving average alternative hypothesis $\theta_S < 1$ is given by

$$
L_{\text{MA}} = \frac{1}{T_\tau} \frac{\Delta_S Y'[\Omega_Y]^{-2} \Delta_S Y}{\Delta_S Y'[\Omega_Y]^{-1} \Delta_S Y}. \tag{2.49}
$$

The null hypothesis is rejected for large values of L_{MA}.

A second, computationally simpler representation of this test statistic is also considered by Tam and Reinsel. This effectively exploits the equivalence between deterministic seasonality and the seasonally differenced process with $\theta_S = 1$, as pointed out by Bell (1987) and discussed above in Subsection 2.3.3. Thus, instead of consideration of the seasonal MA representation of (2.48), the equivalent levels representation

$$
y_t = \sum_{s=1}^{S} \gamma_s \delta_{st} + \varepsilon_t, \qquad t = -S + 1, \ldots, 1, \ldots, T \tag{2.50}
$$

can be used. Note that in the latter, the period begins at year $\tau = 0$ to include the starting values that are important for the equivalence of Bell. Applying OLS to all $T + S$ observations of (2.50) yields residuals $\widehat{\varepsilon}_t = \widehat{\varepsilon}_{s\tau}$, and the test statistic can be calculated as

$$
L_{\text{MA}} = \frac{1}{ST_\tau(T_\tau + 1)\widetilde{\sigma}^2} \sum_{s=1}^{S} \sum_{\tau=0}^{T_\tau} \left[\sum_{j=0}^{\tau} \widehat{\varepsilon}_{sj} \right]^2,
$$

$$
= \frac{1}{ST_\tau(T_\tau + 1)\widetilde{\sigma}^2} \sum_{\tau=0}^{T_\tau} (\widehat{E}^a_\tau)' \widehat{E}^a_\tau, \tag{2.51}
$$

where $\widetilde{\sigma}^2$ is again the OLS estimator of σ^2 and $\widehat{E}^a_\tau$ is the vector of accumulated season-specific residuals at the end of year τ. Thus, $\widehat{E}^a_\tau$ is constructed in the same way as $\widehat{E}^a_t$ above. However, only end of year values are considered for the former, so the correspondence applies for $t = S\tau$ over $\tau = 0, 1, \ldots, T_\tau$.

INVOICE

Invoice #	611234	
Customer #	192988	

Ship From:

LABYRINTH BOOKS SALES ANNEX
358 SAW MILL RIVER ROAD
MILLWOOD BUSINESS CENTER
MILLWOOD, NY 10546
USA

Fax:(914) 762-6261

Bill To:

ROY BLACK
3861 NEW SALEM
OKEMOS, MI 48864

Contact: ROY BLACK

Ship To:

ROY BLACK
3861 NEW SALEM
OKEMOS, MI 48864

Contact: ROY BLACK

Ship Via		Terms	Ordered By	Salesperson
U.S. POSTAL SERVICE		Amazon Payments	AMZ	AMZ

					PO #
					104-4284577-7350628

Ship Date	Order Date	Location	Item #	Item Description	Price	Extended Price
07/14/09	07/14/09	A23C	053948	The Econometric Analysis of Seasonal Time Series (Them	4.98	4.98

Order Quantity	Shipped Quantity
1	1

1 Total # of Units

Your order has been paid by Amazon Pymts.
Please be sure to visit our "Sales Annex"
at www.Labyrinthbooks.com for thousands
of titles all at 40%-90% off retail.

011234

Total Paid		8.97
Balance Due		0.00
Page #	1	

Subtotal	4.98
Freight	3.99
Invoice Total	8.97

When arguments corresponding to those of Subsection 2.4.1 are used, then for the vectors E_τ^a of accumulated true disturbances ε_t,

$$\frac{1}{\sigma^2(T_\tau + 1)^2} \sum_{\tau=0}^{T_\tau} \left(E_\tau^a\right)' E_\tau^a \Rightarrow \int_0^1 W_\varepsilon(r)' W_\varepsilon(r)\, dr, \qquad (2.52)$$

where $W_\varepsilon(r)$ is an $S \times 1$ vector of independent standard Brownian motions. Since $E_\tau^a = (\sum_{j=1}^\tau \varepsilon_{1j}, \dots, \sum_{j=1}^\tau \varepsilon_{Sj})'$, the sth element of $W_\varepsilon(r)$ derives through the accumulation of the i.i.d. disturbances relating to season s alone. The use of residuals in (2.51), which have zero mean for season $s = 1, \dots, S$, rather than the true disturbances implies that

$$L_{\mathrm{MA}} \Rightarrow \frac{1}{S} \int_0^1 [W_\varepsilon(r) - r W_\varepsilon(1)]' [W_\varepsilon(r) - r W_\varepsilon(1)]\, dr. \qquad (2.53)$$

The scaling by $T_\tau(T_\tau + 1)$ rather than T_τ^2 and the use of $\tilde{\sigma}^2$ rather than σ^2 have asymptotically negligible effects. Thus, although not referred to by Tam and Reinsel as the von Mises distribution, it can be seen that $L_{\mathrm{MA}} \Rightarrow \frac{1}{s} VM(S)$.

Tam and Reinsel (1997) extend their analysis to allow the disturbances in (2.47) to follow a stationary and invertible ARMA process, $z_t = \phi(L)^{-1}\theta^*(L)\varepsilon$, so that the model used becomes

$$\phi(L)\Delta_S y_t = \theta(L)\varepsilon_t = \theta^*(L)(1 - \theta_S L^S)\varepsilon_t, \qquad (2.54)$$

with the null hypothesis again $\theta_S = 1$. As in the Canova–Hansen case, an estimator is required that is consistent under both the null and alternative hypotheses. The asymptotic distribution is then unaffected by the use of a parametric correction based on the estimates of $\phi(L)$ and $\theta^*(L)$. Tam and Reinsel (1998) further extend the approach to include an intercept in (2.54), implying that an underlying deterministic trend (common to all seasons) may be present in the DGP for y_t. Issues related to the role of intercepts in the seasonal random walk are detailed in Chapter 3. However, for the present purposes, we note that this intercept has nonnegligible effects on the asymptotic distribution of the test statistic and different critical values are required.

2.4.4 Some Comments

It is obvious that the Canova–Hansen–Caner test of deterministic seasonality is related to the test of Tam and Reinsel. Indeed, in our exposition of the former, where we include the overall mean μ in the test, the tests follow the same asymptotic distribution aside from a scaling by S. Not surprisingly, therefore, the tabulated asymptotic critical values in Canova and Hansen (1995) and in Tam and Reinsel (1997) are very similar when this scaling is taken into account.

Such critical values are in practice obtained from Monte Carlo simulations, and this might well account for the differences.

The relationship between the test statistics is, however, deeper than the fact that they follow the same asymptotic distribution. Putting aside for the moment the different ways in which the residuals are calculated, one sees that (2.35) and (2.51) are very similar. Indeed, in the uncorrelated case (where $\widehat{Z}_t^a = \widehat{E}_t^a$ and $\widehat{\Omega}_Z = \widetilde{\sigma}^2 I_S$), when the deterministic seasonality model is estimated over the same sample period in both cases, then the statistics will be identical except for the scaling. Although there appears to be a difference between them in that the Canova–Hansen statistic sums over all observations and the Tam–Reinsel statistic sums over years, this distinction can be seen to be irrelevant once it is recognized that disturbances (and hence residuals) for season s occur only once a year. In this uncorrelated case, and with estimation undertaken in each case over years $\tau = 1, \ldots, T_\tau$, then it can be observed that the precise relationship between these statistics is

$$L_{\text{MA}} = \frac{T_\tau}{S(T_\tau - 1)} L. \tag{2.55}$$

The relationship of (2.55) applies to the overall test statistic for the null hypothesis that the parameters $\gamma_1, \ldots, \gamma_S$ of the deterministic component of (2.1) are constant over time. The advantage of the Canova–Hansen–Caner approach over that of Tam and Reinsel, however, is that a test can be applied for deterministic seasonality without constancy of the overall mean being part of the test. Further, Caner's statistic has a decomposition that permits each of the seasonal frequencies to be examined separately; this is clearly useful in allowing a departure from a deterministic seasonal process to be investigated in terms of the seasonal frequencies. On the other hand, however, the Tam–Reinsel approach effectively builds up the overall test statistic by investigating each of the seasons. Thus, this form would be useful for an investigation of whether departure from the null hypothesis process of constant deterministic components is associated with, say, only one or two specific seasons. Therefore, the form of the test statistic to be used should depend on the issues of interest in a specific case.

There are also important practical differences between the approaches to estimation adopted in Canova and Hansen (1995), Caner (1998), and Tam and Reinsel (1997). As noted above, the first and third of these papers advocate direct estimation of the deterministic seasonality model in the levels of y_t, whereas the approach of the second paper is to estimate the (theoretically equivalent) model in seasonal differences. Indeed, it might also be mentioned that Tam and Reinsel propose the use of this latter specification in the case in which the disturbances may follow an ARMA process. Under the null hypothesis, this annual difference specification has a noninvertible moving average disturbance. Although these authors include some discussion of estimation in the circumstances in which the MA process may be noninvertible and their conclusions are reassuring, this could remain an important issue in practice. For example, Newbold, Agiakloglou, and Miller (1994) find that results obtained for estimation of ARMA models can change with different approaches to MA estimation and, indeed, with different computer packages apparently adopting the same approach. In general, exact maximum likelihood is preferred for MA processes because of its better properties near the invertibility boundary [Ansley and Newbold (1980) and Osborn (1982)].

Another important difference is that Canova and Hansen (1995) adopt a nonparametric approach, whereas the other authors adopt parametric ones. The Monte Carlo study of Caner indicates that the parametric approach yields a test with better properties in this context, but independent confirmation of this would be valuable. Indeed, since the test statistics proposed by Canova and Hansen (1995), Caner (1998), and Tam and Reinsel (1997) are effectively the same, we believe that the issues to be settled in this context are now practical ones concerned with model specification and estimation (parametric versus nonparametric, levels versus seasonal difference forms).

As we have emphasized in our discussion throughout this chapter, the competing hypotheses generally examined for the nature of seasonality are that it is either deterministic or of a nonstationary stochastic form. We have here considered the former in some detail and examined tests of the null hypothesis that deterministic seasonality is appropriate for an observed time series. However, despite their complementarity, practitioners generally start from the null hypothesis position that seasonality is of the nonstationary stochastic form. This is discussed in the next chapter.

3 Seasonal Unit Root Processes

3.1 Introduction

The two previous chapters have introduced seasonal unit root processes and considered some of their implications. In particular, Chapter 2 discussed the special case of a seasonal random walk process, showing it to consist of S independent random walk processes with one of these processes attached to each of the seasons of the year.

It is now well known that a series generated by a unit root process can wander widely and smoothly over time without any inherent tendency to return to its underlying mean value; see, for example, Granger (1986). In the seasonal context, there are S unit root processes, none of which has an inherent tendency to return to a deterministic pattern. As a result, the values for the seasons can wander widely and smoothly in relation to each other, making any underlying relationship between the expected values for the different seasons effectively irrelevant in practice. It is often remarked that, in the presence of seasonal unit roots, summer can become winter.

The immediately following section explores the implications of seasonal unit root processes in more detail, before the discussion moves on to consider tests of the unit root hypothesis in Sections 3.3–3.5. The analysis is extended to seasonal cointegration in Section 3.6. Some of the discussion of this chapter is quite technical, which is the nature of material dealing with the asymptotics of nonstationary processes. Our aim, however, is to make the material as accessible as possible and to draw together strands of recent research in this area. As was found in the discussion of tests of deterministic seasonality in Chapter 2, some approaches that initially appear to be different will be shown to be closely related. The final section of this chapter contains some reflections on the empirical results obtained to date in relation to seasonal unit root processes.

3.2 Properties of Seasonal Unit Root Processes

Stochastic nonstationary seasonality has already been considered in Chapters 1 and 2. Here we will extend that discussion and we will often refer to *seasonal integration*. Formally,

Definition 3.1: The nonstationary stochastic process y_t, observed at S equally spaced time intervals per year, is said to be seasonally integrated of order d, denoted $y_t \sim SI(d)$, if $\Delta_S^d y_t$ is a stationary, invertible ARMA process.

The first-order annual differencing operator Δ_S is often called the seasonal differencing filter. The definition implies that if annual differencing renders y_t a stationary and invertible process, then $y_t \sim SI(1)$. In practice, orders $d > 1$ for seasonal integration are rarely considered. It may be recalled that in Chapter 2 we discussed the equivalence between deterministic seasonality and a specific form of a process involving seasonal differencing. It should be noted that this equivalence does not imply that the deterministic seasonal process is seasonally integrated. This is because seasonally differencing such a process induces the factor Δ_S in the moving average polynomial, thereby causing noninvertibility. Thus, a deterministic seasonal process and a seasonally integrated one are distinct processes that cannot be equivalent.

The simplest case of a seasonally integrated process is the seasonal random walk considered in previous chapters. Here we extend our earlier discussion in various ways, in particular by allowing the process to include drift terms and additional stationary dynamics. Later, in Subsection 3.2.4, we also consider the relationship between representations of the seasonal random walk as analogous to that between the dummy variable and trigonometric representations of deterministic seasonality compared in Chapter 2. Finally, this section contains a summary of the important distributional building blocks for the derivation of the asymptotic distributions of the seasonal unit root test statistics considered later in the chapter.

3.2.1   The Seasonal Random Walk with Constant Drift

The seasonal random walk discussed in earlier chapters is not very realistic in that it makes no allowance for the trend behavior typically observed in economic time series. To generalize the discussion of Subsection 2.3.2 of Chapter 2, consider the simple extension of adding a constant drift term γ, so that

$$\Delta_S y_t = \gamma + \varepsilon_t, \qquad t = 1, 2, \ldots, T, \tag{3.1}$$

with $\gamma \neq 0$ and $\varepsilon_t \sim$ i.i.d. $(0, \sigma^2)$.

As noted in Chapter 2, the seasonal random walk process is, in effect, S separate random walk processes, one relating to each of the S seasons. This implication, which is used by Dickey, Hasza, and Fuller (1984) and emphasized by Osborn (1993), follows because each process $y_{s\tau}$ ($s = 1, \ldots, S$) is influenced only by disturbances relating to the specific season s. This is also true in (3.1), which, by repeated substitution, can be seen to imply

$$y_{s\tau} = \gamma\tau + y_{s0} + \sum_{j=1}^{\tau} \varepsilon_{sj}, \qquad s = 1, \ldots, S, \ \tau = 1, \ldots, T_\tau, \quad (3.2)$$

when the double subscript notation is used. Further, from (3.2), the vector representation for $Y_\tau = (y_{1\tau}, \ldots, y_{S\tau})'$, namely for the S observations relating to year τ, is easily seen to be

$$Y_\tau = \gamma\tau e_S + Y_0 + \sum_{j=1}^{\tau} U_j, \qquad \tau = 1, \ldots, T_\tau, \quad (3.3)$$

where e_S is an $S \times 1$ vector with all elements equal to unity and the disturbance vector for year τ is $U_\tau = (\varepsilon_{1\tau}, \ldots, \varepsilon_{S\tau})'$. From the above assumptions, $U_\tau \sim$ i.i.d. $(0, \sigma^2 I_S)$. These last two equations can be viewed as generalizations of the conventional nonseasonal random walk with drift, $\Delta y_t = \gamma + \varepsilon_t$, which implies

$$y_t = \gamma t + y_0 + \sum_{j=1}^{t} \varepsilon_j, \qquad t = 1, 2, \ldots, T. \quad (3.4)$$

Two other features of (3.2) are worthy of comment. First, the unconditional mean is

$$E(y_{s\tau}) = \gamma\tau + E(y_{s0}). \quad (3.5)$$

Thus, although the process (3.1) does not explicitly contain deterministic seasonal effects, these are implicitly included when $E(y_{s0})$ varies over $s = 1, \ldots, S$. Second, it is apparent that (3.2) defines a random walk process with drift, giving rise to the same property of a smooth and wide deviation from trends discussed by Granger (1986) for the nonseasonal case. While the random walk involves the disturbances for the specific season s only, (3.5) indicates that the slope of the trends for $y_{s\tau}$ is constant at γ units per year irrespective of the season s.

One consequence of (3.1) is that any linear combination of the separate random walk processes for $s = 1, \ldots, S$ can itself be represented as a random walk. For example, the annual process defined as the change in y_t between the first and third quarters of each year is $y_{3\tau} - y_{1\tau}$ ($\tau = 1, \ldots, T_\tau$), and from (3.2),

$$y_{3\tau} - y_{1\tau} = (y_{30} - y_{10}) + \sum_{j=1}^{\tau}(\varepsilon_{3j} - \varepsilon_{1j}),$$

$$= (y_{30} - y_{10}) + \sum_{j=1}^{\tau}\eta_j, \quad (3.6)$$

where $\eta_j = \varepsilon_{3j} - \varepsilon_{1j} \sim$ i.i.d. $(0, 2\sigma^2)$. This new random walk has constant mean $E(y_{30}) - E(y_{10})$ but no trend. The accumulation of disturbances η_j allows the differences $y_{3\tau} - y_{1\tau}$ to wander far from this mean over time, giving rise to the phenomenon already referred to as, "summer can become winter."

3.2.2 The Seasonal Random Walk with Seasonally Varying Drift

From the discussion of the previous subsection, it is relatively straightforward to see the implications of allowing the drift term in (3.1) to be seasonally varying. The process then becomes

$$\Delta_S y_{s\tau} = \gamma_s + \varepsilon_{s\tau}, \qquad s = 1, \ldots, S, \ \tau = 1, \ldots, T_\tau, \tag{3.7}$$

or, corresponding to (3.2),

$$y_{s\tau} = y_{s0} + \gamma_s \tau + \sum_{j=1}^{\tau} \varepsilon_{sj}. \tag{3.8}$$

Taking expectations, this last equation can be seen to imply that the mean of the process for period t is $E(y_t) = E(y_{s\tau}) = E(y_{s0}) + \gamma_s \tau$, which (in general) is a seasonally varying deterministic trend in both the intercept and the slope. Indeed, (3.8) contains no link between the observations for the S seasons, since they follow different underlying trends with the deviations from these trends being independent random walk processes. As already noted in the context of deterministic seasonality in Chapter 2, it may be economically implausible to assume that the observations follow different deterministic trends over the seasons of the year since this implies that the seasons will diverge in their trends. Put slightly differently, analogous to (3.6), the seasonally varying drift process implies that

$$y_{3\tau} - y_{1\tau} = (y_{30} - y_{10}) + (\gamma_3 - \gamma_1)\tau + \sum_{j=1}^{\tau}(\varepsilon_{3j} - \varepsilon_{1j}), \tag{3.9}$$

which has a nonzero trend component with slope $\gamma_3 - \gamma_1$. Because of the implausibility of such behavior in many circumstances, we generally assume in what follows that any drift in a seasonally integrated process is constant over the seasons.

3.2.3 More General Stochastic Processes

To generalize the above discussion, weakly stationary autocorrelation can be permitted in the $SI(1)$ process. That is, the constant drift process (3.1) can be generalized to the seasonally integrated ARMA process with drift:

$$\phi(L)\Delta_S y_t = \gamma + \theta(L)\varepsilon_t, \qquad t = 1, 2, \ldots, T, \tag{3.10}$$

where, as before, $\varepsilon_t \sim$ i.i.d. $(0, \sigma^2)$, while the polynomials $\phi(L)$ and $\theta(L)$ have all roots outside the unit circle. It is, of course, permissiable that these polynomials take the multiplicative form of the seasonal ARMA model of Box and Jenkins (1976) with, for example, $\phi(L) = \phi_1(L)\phi_S(L^S)$ where $\phi_1(L)$ is a polynomial in L and $\phi_S(L^S)$ is a polynomial in L^S.

Inverting the stationary autoregressive polynomial and defining $z_t = \phi(L)^{-1}\theta(L)\varepsilon_t$, we can write (3.10) as

$$\Delta_S y_t = \gamma^* + z_t, \qquad t = 1, \ldots, T, \tag{3.11}$$

where $\gamma^* = \gamma/\phi(1) = E(\Delta_S y_t)$. The process superficially looks like the seasonal random walk with drift, namely (3.1). Writing in double subscript notation and substituting for lagged z leads to the corresponding result; that is,

$$y_{s\tau} = \gamma^*\tau + y_{s0} + \sum_{j=1}^{\tau} z_{sj}, \qquad s = 1, \ldots, S, \ \tau = 1, \ldots, T_\tau. \tag{3.12}$$

The characteristic of the mean following the same trend over the S seasons, but with intercepts varying with $E(y_{s0})$, applies in this case as before. Also, as in (3.2), the stochastic component of (3.12) implies that there are S distinct unit root processes of the form $\sum_{j=1}^{\tau} z_{sj}$, one corresponding to each of the S seasons. The important distinction is that the processes here may be autocorrelated and cross correlated. Nevertheless, it is only the stationary components z_{sj} that are correlated.

Using the observation and (weakly stationary) disturbance vectors for year τ, Y_τ and Z_τ respectively, we find the vector representation of (3.11) is

$$Y_\tau = Y_{\tau-1} + \gamma^* e_S + Z_\tau,$$
$$= (\gamma^*\tau)e_S + Y_0 + \sum_{j=1}^{\tau} Z_j, \tag{3.13}$$

with e_S in (3.13) being an $S \times 1$ vector with each element equal to unity. The disturbances here follow a stationary vector ARMA process:

$$\Phi(L)Z_\tau = \Theta(L)U_\tau. \tag{3.14}$$

Leaving the details until Chapter 6, we find it sufficient to note at this stage that $\Phi(L)$ and $\Theta(L)$ are appropriately defined $S \times S$ polynomial matrices in L with all roots outside the unit circle.

Now, in (3.13) we have a vector ARMA process in $\Delta_S Y_\tau$, which is a vector ARIMA process in Y_τ. In the terminology of Engle and Granger (1987), the S processes in the vector Y_τ cannot be cointegrated if this is the DGP. Expressed in a slightly different way, the seasonally integrated DGP for y_t implies that the vector Y_τ contains S unit roots (one corresponding to each of the seasons). Thus, the implication drawn out from the seasonal random walk of Subsection 3.2.1

that any linear combination of the annual series $y_{s\tau}$ over $s = 1, \ldots, S$ is itself an $I(1)$ process applies to any seasonally integrated process. The economic plausibility of such nonstationarity between the different seasons is, however, a question that remains largely unanswered.

3.2.4 Transformations and Seasonal Unit Roots

Chapter 2, and especially Section 2.2, discusses the one-to-one relationship between the dummy variable and the trigonometric representations of deterministic seasonality, with this relationship being captured by the nonsingular matrix R. More specifically, premultiplication of the time-domain dummy variable representation by R' effectively converts the representation to the frequency-domain trigonometric one (except for the scaling terms d_i). This transformation can be applied to a sequence of observations on y_t.

Consider first the annual observation vector for year τ, Y_τ, and define

$$Y_\tau^{(S)} = R'Y_\tau. \tag{3.15}$$

This transformation applied to Y_τ has already been met in the context of the Canova–Hansen test in Chapter 2, where it was used under the null hypothesis to yield terms associated with each of the coefficients of the trigonometric representation of deterministic seasonality. The transformation of (3.15) can also be applied to any sequence of S consecutive observations ending in some period t. As we shall see, this is helpful in enabling the separate unit roots implied by $\Delta_S = (1 - L^S)$ to be considered one at a time.

Once again, we illustrate this by considering the quarterly case, where we define the variables $y_t^{(1)}$, $y_t^{(2)}$, $y_t^{(3)}$, and $y_t^{(4)}$ by

$$\begin{bmatrix} y_t^{(1)} \\ y_t^{(2)} \\ y_t^{(3)} \\ y_t^{(4)} \end{bmatrix} = \begin{bmatrix} 1 & 1 & 1 & 1 \\ -1 & 1 & -1 & 1 \\ 0 & -1 & 0 & 1 \\ 1 & 0 & -1 & 0 \end{bmatrix} \begin{bmatrix} y_{t-3} \\ y_{t-2} \\ y_{t-1} \\ y_t \end{bmatrix}. \tag{3.16}$$

Notice that the matrix implicitly defined as R' in (3.16) has a different order of rows from R' as defined from (2.9) in Chapter 2. This is, however, innocuous because it simply relates to the numbering of the $y_t^{(i)}$. The reason we use the form of (3.16) is to relate the variables defined here to those used in the seasonal unit root literature and especially in the paper by Hylleberg, Engle, Granger, and Yoo (1990), referred to as HEGY. It might also be noted that $y_t^{(4)}$ is redundant in the sense that $y_t^{(4)} = -Ly_t^{(3)}$.

Now, since

$$1 - L^4 = (1 - L)(1 + L)(1 + L^2), \tag{3.17}$$

the quarterly $SI(1)$ process with drift $\Delta_4 y_t = \gamma + z_t$ can be seen to imply

$$(1 - L)y_t^{(1)} = \gamma + z_t, \tag{3.18}$$

$$(1 + L)y_t^{(2)} = \gamma + z_t, \tag{3.19}$$

$$(1 + L^2)y_t^{(3)} = \gamma + z_t. \tag{3.20}$$

The processes of (3.18)–(3.20) result when it is appreciated that the variables defined in (3.16) each impose two of the three nonstationary real factors $(1 - L), (1 + L)$, and $(1 + L^2)$ of $1 - L^4$. In each case, the remaining factor yields the nonstationary dynamic properties of the process. In particular, the dynamics of $y_t^{(1)}$ include the usual unit root of 1, and those of $y_t^{(2)}$ include the unit root -1; the stochastic properties of $y_t^{(3)}$ are determined by the complex unit roots $\pm i$ that arise from the factor $(1 + L^2)$. It will also be recalled from Chapter 2 that the unit root 1 is associated with the zero spectral frequency, while the seasonal unit roots of -1 and $\pm i$ are associated with the frequencies π and $\pi/2$, respectively.

Examined in a little more detail, (3.18) implies

$$y_t^{(1)} = y_0^{(1)} + \gamma t + \sum_{j=0}^{t-1} z_{t-j}, \tag{3.21}$$

as $y_t^{(1)}$ is a conventional unit root process with drift. Turning to $y_t^{(2)}$, we should note that (3.19) implies a "bounce back" as

$$y_t^{(2)} = -y_{t-1}^{(2)} + \gamma + z_t, \tag{3.22}$$

and hence $y_t^{(2)}$ exhibits a half-cycle every period and a full cycle every two periods (a semiannual cycle for quarterly data). Substituting for lagged $y_t^{(2)}$ in (3.22), we have

$$y_t^{(2)} = \begin{cases} -y_0^{(2)} + \gamma + \sum_{j=0}^{t-1}(-1)^j z_{t-j}, & t \text{ odd} \\ y_0^{(2)} + \sum_{j=0}^{t-1}(-1)^j z_{t-j}, & t \text{ even} \end{cases}. \tag{3.23}$$

One property of note for the process $y_t^{(2)}$ is that the constant γ does not contribute any trend component. Further, while the stationary z_t accumulate over time, the accumulation involves the sign alternating each period.

Finally, from (3.20),

$$y_t^{(3)} = -y_{t-2}^{(3)} + \gamma + z_t. \tag{3.24}$$

This process implies a bounce back after two periods and a full cycle after four (an annual cycle when $S = 4$). Also, because y_{t-2}, but not y_{t-1}, influences

y_t, then there must in fact be *two* distinct unit root processes, one linking the odd-numbered observations ($t = 1, 3, 5, \ldots$) and the other linking the even-numbered observations ($t = 2, 4, 6, \ldots$). One way of capturing these two separate processes is by using $y_t^{(3)}$ and $y_t^{(4)} = -Ly_t^{(3)}$. Substituting in (3.24) gives, for even t,

$$
y_t^{(3)} = \begin{cases} -y_0^{(3)} + \gamma + \sum_{j=0}^{g}(-1)^j z_{t-2j}, & t = 2, 6, 10, \ldots \\[2mm] y_0^{(3)} + \sum_{j=0}^{g}(-1)^j z_{t-2j}, & t = 4, 8, 12, \ldots \end{cases} , \qquad (3.25)
$$

where $g = \text{int}[(t - 1)/2]$. The two expressions in (3.25) apply also for $t = 1, 5, 9, \ldots$ and $t = 3, 7, 11, \ldots$, respectively, except that the starting value $y_0^{(3)}$ becomes $y_{-1}^{(3)}$. Also note that, in contrast to the underlying linear trend for $y_t^{(1)}$ implied by a nonzero γ, the process $y_t^{(3)}$, like $y_t^{(2)}$, implies no overall trend.

Beaulieu and Miron (1993) generalize the HEGY definitions of (3.16) for the case of monthly data. Indeed, the variables Beaulieu and Miron define can be obtained immediately from the transformation (3.15) applied to a sequence of twelve values ending in period t with R appropriately defined for the monthly case (see Section 2.2). Once again, the transformations isolate each of the nonstationary unit roots relating to the zero and seasonal frequencies for the nonstatonary polynomial $(1 - L^{12})$. The eleven seasonal unit roots appear as five pairs of complex roots on the unit circle, together with the real root -1.

One of the themes of Subsections 3.2.1–3.2.3 is that seasonal integration implies the presence of a nonstationary unit root process associated with each season. Dickey, Hasza, and Fuller (1984), or DHF, appear to have been the first to recognize this implication. Much of the subsequent seasonal unit root literature, however, seeks to examine the unit roots implied by the factorization of Δ_S, rather than explicitly examining the unit roots associated with each of the seasons. It is, therefore, useful to consider the relationship between these two approaches.

For simplicity of exposition, let us assume there are precisely τ years of observations available to time t and that the starting values of the nonstationary process for y_t are fixed. For year τ, the stochastic nonstationarity in Y_τ is, from (3.13), given by $\sum_{j=1}^{\tau} Z_j$. When $R'Y_\tau$ is examined through the HEGY approach, the relevant stochastic process is $R' \sum_{j=1}^{\tau} Z_j$. Since R is nonsingular, there is a one-to-one relationship between the stochastic processes $\sum_{j=1}^{\tau} Z_j$ and $R' \sum_{j=1}^{\tau} Z_j$. One implication is that the insight of DHF carries over, so that the S unit roots associated with the seasons in a seasonally integrated process and the S unit roots considered in the HEGY approach are different manifestations of precisely the same phenomenon. In other words, they are equivalent. Although this equivalence is analogous to the equivalence between the dummy variable and trigonometric representations of deterministic seasonality (see Chapter 2), it seems to be frequently overlooked in the seasonal unit root literature.

However, there is also one difference that offers a practical advantage to the HEGY approach in some contexts. That is, the variables defined by HEGY in the quarterly case, and by Beaulieu and Miron in the monthly one, are asymptotically orthogonal in that if the observations on these constructed variables over $t = 1, \ldots, T$ are used to define the columns of a $T \times S$ matrix X, then asymptotically $(1/T^2)X'X$ is diagonal. This asymptotic orthogonality is partly a consequence of the orthogonality of the matrix R, discussed in Chapter 2. No corresponding orthogonality applies if the original S variables of Y_τ are considered. In the seasonal random walk case the variables are mutually independent, but with unit root processes the corresponding $S \times S$ cross-product matrix

$$\frac{1}{T_\tau^2} \sum_{\tau=1}^{T_\tau} Y_\tau Y_\tau' \qquad (3.26)$$

is not asymptotically diagonal. This is discussed in the next subsection.

3.2.5 Asymptotic Distributions

In deriving the asymptotic distributions of the test statistics for seasonal unit roots, we will use various results concerning the asymptotic properties of integrated processes. Although some of these properties have already been used in Chapter 2, for convenience we repeat these here and hence summarize all the important asymptotic distributional results used in subsequent sections of the present chapter.

The basic results concerning the asymptotic distributions for integrated processes derive from the driftless random walk case of $y_t = y_{t-1} + \varepsilon_t$ with $y_0 = 0$. With T observations ($t = 1, \ldots, T$) available on this process, Wiener distribution theory [see, e.g., Hamilton (1994a) or Banerjee et al. (1993)] implies that, as $T \to \infty$,

$$T^{-1/2} y_t \Rightarrow \sigma W(r), \qquad (3.27)$$

where $W(r)$ is standard Brownian motion (also known as a Wiener process), derived from an i.i.d. $(0, 1)$ disturbance, and the right double arrow indicates convergence in distribution. Further,

$$T^{-3/2} \sum_{t=1}^{T} y_t \Rightarrow \sigma \int_0^1 W(r)\,dr, \qquad (3.28)$$

$$T^{-1} \sum_{t=1}^{T} \varepsilon_t y_{t-1} \Rightarrow \sigma^2 \int_0^1 W(r)\,dW(r), \qquad (3.29)$$

$$T^{-2} \sum_{t=1}^{T} y_t^2 \Rightarrow \sigma^2 \int_0^1 [W(r)]^2\,dr. \qquad (3.30)$$

Combining (3.27) and (3.28) implies that when the sample mean $\bar{y} = T^{-1} \sum_{t=1}^{T} y_t$ is subtracted from y_t, then

$$T^{-1/2}(y_t - \bar{y}) \Rightarrow \sigma \left[W(r) - \int_0^1 W(r)\,dr \right]. \tag{3.31}$$

In a fairly obvious terminology, $\widetilde{W}(r) = W(r) - \int_0^1 W(r)dr$ is often referred to as demeaned Brownian motion.

Moving on to consider the role of seasonality in nonstationary processes, we have emphasized in our discussion that the seasonal random walk consists of S random walk processes. Thus, the basic distributional results for seasonal unit root tests can be derived from those of the processes $y_{s\tau} = y_{s,\tau-1} + \varepsilon_{s\tau}$ with $y_{s0} = 0\,(s = 1, \ldots, S)$. We continue to assume observations on each process for $\tau = 1, \ldots, T_\tau$ while we also assume that $\varepsilon_{s\tau} \sim$ i.i.d. $(0, \sigma^2)$. Note the common variance assumption across the processes. With T replaced by T_τ and with $W(r)$ replaced by the standard Brownian motion for season s, denoted by $W_s(r)$, the results of (3.27)–(3.30) apply immediately for each of the S random walk processes. To modify (3.31), we need to note that the relevant sample mean is the season specific one $\bar{y}_s = T_\tau^{-1} \sum_{\tau=1}^{T_\tau} y_{s\tau}$ and then

$$T_\tau^{-1/2}(y_{s\tau} - \bar{y}_s) \Rightarrow \sigma \widetilde{W}_s(r), \qquad s = 1, \ldots, S, \tag{3.32}$$

where the demeaned Brownian motion is $\widetilde{W}_s(r) = W_s(r) - \int_0^1 W_s(r)dr$. We also need to emphasize, when considering the modifications to take account of the S processes $y_{s\tau}$, that the S Brownian motions $W_s(r)$ are independent. This follows since $\varepsilon_{s\tau}$ (and hence $y_{s\tau}$) are independent over seasons $s = 1, \ldots, S$.

We will also be interested in asymptotic distributional properties concerned with the vector Y_τ. In particular, we will require generalizations of (3.29) and (3.30) to deal with this case. These generalizations are

$$T_\tau^{-1} \sum_{\tau=1}^{T_\tau} \varepsilon_{s\tau} y_{q,\tau-1} \Rightarrow \sigma^2 \int_0^1 W_q(r)\,dW_s(r), \qquad q, s = 1, \ldots, S, \tag{3.33}$$

$$T_\tau^{-2} \sum_{\tau=1}^{T_\tau} y_{s\tau} y_{q\tau} \Rightarrow \sigma^2 \int_0^1 W_s(r)W_q(r)\,dr, \qquad q, s = 1, \ldots, S, \tag{3.34}$$

respectively [see, e.g., Hamilton (1994a)]. It is especially important to note that when $q \neq s$, despite the fact that $y_{q\tau}$ and $y_{s\tau}$ are independent, (3.34) implies that $T_\tau^{-2} \sum_{\tau=1}^{T_\tau} y_{s\tau} y_{q\tau}$ converges to a distribution and not to zero. This is the result that leads to the matrix of (3.26) not being diagonal. Indeed, in a slightly different context, such nondiagonality leads to the "spurious regression" phenomenon analyzed by Phillips (1986).

3.3 Testing Seasonal Integration

In this section, we discuss a number of procedures proposed to test the null hypothesis of seasonal integration. To keep the analysis as simple as possible, we will focus primarily on the seasonal random walk case of (3.1), with the additional assumptions that all starting values ($y_{-s+1}, \ldots, y_0$) and the drift, γ, are zero. While recognizing that such assumptions are unrealistic, we begin by making them in order to clarify the basic distributional results concerning seasonal unit root test statistics. Similarly, to clarify the exposition, we also begin by assuming that augmentation of the test regression to account for autocorrelation is unnecessary. It is vitally important to appreciate that some of these assumptions play a nontrivial role, so that the extensions discussed in Section 3.4 below do need to be addressed in practice.

3.3.1 Dickey–Hasza–Fuller Test

The first test of the null hypothesis $y_t \sim SI(1)$ was proposed by Dickey, Hasza, and Fuller (or DHF) (1984) as a direct generalization of the test proposed by Dickey and Fuller (1979) for a nonseasonal AR(1) process. Assuming that the process is known to be the seasonal AR(1),

$$y_t = \phi_S y_{t-s} + \varepsilon_t, \qquad t = 1, 2, \ldots, T, \tag{3.35}$$

with $\varepsilon_t \sim$ i.i.d. $(0, \sigma^2)$, then DHF proposed a test of $\phi_S = 1$ in (3.35). The alternative that y_t is stationary implies the one-sided alternative hypothesis $\phi_S < 1$. An equivalent, but often more convenient, way to set up the problem is to reparameterize (3.35) as

$$\Delta_S y_t = \alpha_S y_{t-s} + \varepsilon_t; \tag{3.36}$$

where $\alpha_S \equiv -(1 - \phi_S)$. The null hypothesis of seasonal integration corresponds to $\alpha_S = 0$, while the alternative of a stationary stochastic seasonal process implies $\alpha_S < 0$.

In common with the usual Dickey–Fuller approach, the test is typically implemented as the "t ratio" for the estimated value of α_S in (3.36). This is given by

$$t(\widehat{\alpha}_S) = (\widehat{\alpha}_S)/[\widehat{se}(\widehat{\alpha}_S)] = \left(\sum_{t=1}^{T} y_{t-s} \Delta_S y_t \right) \bigg/ \left\{ \widetilde{\sigma} \left[\sum_{t=1}^{T} y_{t-s}^2 \right]^{1/2} \right\},$$

where $\widehat{\alpha}_S$ and $\widehat{se}(\widehat{\alpha}_S)$ are defined by formulas identical to those used in the conventional application of ordinary least squares to (3.36) to estimate α_S and its standard error respectively, while $\widetilde{\sigma}$ is the usual degrees of freedom corrected estimator of σ. Under the null hypothesis $\alpha_S = 0$, the t statistic becomes

$$t(\widehat{\alpha}_S) = \left(\sum_{t=1}^{T} y_{t-S}\varepsilon_t \right) \bigg/ \left\{ \widetilde{\sigma} \left[\sum_{t=1}^{T} y_{t-S}^2 \right]^{1/2} \right\}. \tag{3.37}$$

On the basis of Monte Carlo simulations, DHF tabulated critical values of $t(\widehat{\alpha}_S)$ for various T and S.

In Subsection 3.2.1, we have already noted that DHF refer to the S series that we denote as $y_{s\tau}$ ($s = 1, \ldots, S$). Using this double subscript notation, and continuing to assume for ease of exposition that we have $T = ST_\tau$ sample observations, we can see that (3.37) implies

$$t(\widehat{\alpha}_S) = \frac{\sum_{s=1}^{S} \sum_{\tau=1}^{T_\tau} \varepsilon_{s\tau} y_{s,\tau-1}}{\widetilde{\sigma} \left[\sum_{s=1}^{S} \sum_{\tau=1}^{T_\tau} y_{s,\tau-1}^2 \right]^{1/2}} = \frac{T_\tau^{-1} \sum_{s=1}^{S} \sum_{\tau=1}^{T_\tau} \varepsilon_{s\tau} y_{s,\tau-1}}{\widetilde{\sigma} \left[T_\tau^{-2} \sum_{s=1}^{S} \sum_{\tau=1}^{T_\tau} y_{s,\tau-1}^2 \right]^{1/2}}. \tag{3.38}$$

Asymptotically,

$$\widetilde{\sigma}^2 \overset{p}{\to} \sigma^2, \tag{3.39}$$

where $\overset{p}{\to}$ denotes convergence in probability, while the other terms in (3.38) have convergence properties as stated in Subsection 3.2.5. Substituting those convergence results into (3.38) implies

$$t(\widehat{\alpha}_S) \Rightarrow \left[\sum_{s=1}^{S} \int_0^1 W_s(r) \, dW_s(r) \right] \bigg/ \left\{ \left[\sum_{s=1}^{S} \int_0^1 [W_s(r)]^2 \, dr \right]^{1/2} \right\}. \tag{3.40}$$

The asymptotic distribution of the DHF statistic given by (3.40) is nonstandard, but it is of a similar type to the Dickey–Fuller t distribution. Indeed, it is precisely the Dickey–Fuller t distribution in the special case $S = 1$, when the test regression (3.36) is the usual Dickey–Fuller test regression for an AR(1) process. It can also be seen from (3.40) that the asymptotic distribution for the DHF t statistic depends on S, that is, on the frequency with which observations are made within each year.

To explore the dependence on S a little further, note first that

$$\int_0^1 W_s(r) \, dW_s(r) = \frac{1}{2} \{ [W_s(1)]^2 - 1 \}, \tag{3.41}$$

where $[W_s(1)]^2$ is $\chi^2(1)$ [Banerjee et al. (1993)]. The numerator of (3.40) involves the sum of S such terms that are mutually independent and hence

$$\sum_{s=1}^{S} \int_0^1 W_s(r)\,dW_s(r) = \frac{1}{2}\sum_{s=1}^{S}\{[W_s(1)]^2 - 1\},$$

$$= \frac{1}{2}\{\chi^2(S) - S\}, \tag{3.42}$$

which is half the difference between a $\chi^2(S)$ variable and its mean of S. It has often been observed that the Dickey–Fuller t statistic is not symmetric about zero. Indeed, Fuller (1996) comments that asymptotically the probability of $\widehat{\phi}_1 < 1$ is 0.68 for the nonseasonal random walk because $\Pr[\chi^2(1) < 1] = 0.68$. In terms of (3.40), the denominator is always positive and hence $\Pr[\chi^2(S) < S]$ dictates the probability that $t(\widehat{\alpha}_S)$ is negative. With a seasonal random walk and quarterly data, $\Pr[\chi^2(4) < 4] = 0.59$, while in the monthly case $\Pr[\chi^2(12) < 12] = 0.55$. Therefore, the preponderance of negative test statistics is expected to decrease as S increases. As seen from the percentiles tabulated by DHF, the dispersion of $t(\widehat{\alpha}_S)$ is effectively invariant to S, so that the principal effect of an increasing frequency of observation is a reduction in the asymmetry of this test statistic around zero.

3.3.2 Testing a Unit Root of −1

We need to consider how to test what we might call "nonstandard" unit roots. In particular, as implied by our discussion in Subsection 3.2.4, the approach of testing seasonal integration through the factorization of Δ_S leads us to consider tests for a unit root of −1 and for pairs of complex unit roots. The latter are discussed in the next subsection, with a root of −1 examined in the present subsection. Detailed discussion of the issues related to such tests can be found in Ahtola and Tiao (1987), Chan and Wei (1988), or Chan (1989a).

The case of a unit root of −1 can be considered through the process:

$$y_t^* = -y_{t-1}^* + v_t, \tag{3.43}$$

which, with $y_0^* = 0$, implies

$$y_t^* = \sum_{j=0}^{t-1}(-1)^j v_{t-j}. \tag{3.44}$$

In these equations, we assume that $v_t \sim$ i.i.d. $(0, \sigma^2)$. Now, a test of the unit root could be conducted as a test of the null hypothesis $\alpha^* = 0$ against $\alpha^* > 0$ in

$$(1 + L)y_t^* = \alpha^* y_{t-1}^* + v_t, \qquad t = 1, \ldots, T. \tag{3.45}$$

When estimated by ordinary least squares, the usual t ratio for $\widehat{\alpha}^*$ under the null hypothesis is given by

$$t(\widehat{\alpha}^*) = \left(\sum_{t=1}^{T} v_t y_{t-1}^*\right) \Big/ \left\{\widetilde{\sigma}\left[\sum_{t=1}^{T}(y_{t-1}^*)^2\right]^{1/2}\right\}. \tag{3.46}$$

The asymptotic distributional results in Subsection 3.2.5 cannot be used immediately, because (3.43) is not a random walk process. There is, however, a "mirror image" relationship between the two processes [see, e.g., Fuller (1996) or Chan and Wei (1988)]. Stated simply, this property says that if we consider the process y_t^* of (3.43) and the random walk process $y_t = y_{t-1} + \varepsilon_t$ when $\varepsilon_t = (-1)^t v_t$, then

$$\sum_{t=1}^{T} v_t y_{t-1}^* = -\sum_{t=1}^{T} \varepsilon_t y_{t-1}. \tag{3.47}$$

For readers interested in the details, this property can be established by noting that (3.44) can be used to substitute for y_{t-1}^* and hence, also substituting for v_t in terms of ε_t,

$$\sum_{t=1}^{T} v_t y_{t-1}^* = \sum_{t=1}^{T} v_t \sum_{j=0}^{t-2}(-1)^j v_{t-j-1} = \sum_{t=1}^{T}(-1)^t \varepsilon_t \sum_{j=0}^{t-2}(-1)^j(-1)^{t-j-1}\varepsilon_{t-j-1},$$

$$= -\sum_{t=1}^{T}\varepsilon_t \sum_{j=0}^{t-2}\varepsilon_{t-j-1} = -\sum_{t=1}^{T}\varepsilon_t y_{t-1}.$$

The importance of (3.47) is that, provided that ε_t and v_t are symmetrically distributed around zero, they are identically distributed, and hence $\sum_{t=1}^{T} v_t y_{t-1}^*$ in the context of the process with unit root -1 has identical distributional properties to $-\sum_{t=1}^{T} \varepsilon_t y_{t-1}$ when y_t is a random walk process.

Therefore, when the unit root of -1 is considered (3.29) is replaced by

$$T^{-1}\sum_{t=1}^{T} v_t y_{t-1}^* \Rightarrow -\sigma^2 \int_0^1 W(r)\,dW(r). \tag{3.48}$$

Also note that the variables $(y_t^*)^2 = \left[\sum_{j=0}^{t-1}(-1)^j v_{t-j}\right]^2$ and $(y_t)^2 = \left[\sum_{j=0}^{t-1}\varepsilon_{t-j}\right]^2$ have identical distributional properties since any sign change becomes irrelevant with squaring. Consequently, (3.30) remains valid for a process with unit root -1. When the numerator and denominator of (3.46) are scaled by division by T, then it follows that

$$t(\widehat{\alpha}^*) \Rightarrow -\left[\int_0^1 W(r)\,dW(r)\right]\Big/ \left\{\left[\int_0^1 [W(r)]^2\,dr\right]^{1/2}\right\}, \tag{3.49}$$

which is the mirror image of the usual Dickey–Fuller t distribution. This result is relatively well known and implies that, with a simple change of sign, the Dickey–Fuller tables for the case without a drift can also be used when a unit root of -1 is tested.

3.3.3   Testing Complex Unit Roots

The simplest process that contains a pair of complex unit roots is

$$y_t^* = -y_{t-2}^* + v_t, \tag{3.50}$$

with $y_0^* = y_{-1}^* = 0$, which corresponds to (3.24) with no drift and zero starting values. The disturbance is again assumed to be i.i.d. $(0, \sigma^2)$. As hinted already in the discussion relating to the process $y_t^{(3)}$ in Subsection 3.2.4, the key to our main approach for this complex unit root case is that it can be considered as a seasonal process with $S = 2$ seasons per year. As such, (3.50) implies

$$y_{s\tau}^* = -y_{s,\tau-1}^* + v_{s\tau}, \qquad s = 1, 2, \tag{3.51}$$

$$y_{s\tau}^* = \sum_{j=0}^{\tau-1} (-1)^j v_{s,\tau-j}. \tag{3.52}$$

Note that $y_{s\tau}^*$ $(s = 1, 2)$ are two independent nonstationary processes, one corresponding to each of the two seasons of the year. Each of these processes has the nonstationary properties of the unit root process considered in Subsection 3.3.2.

Analogously to the DHF test, the unit root process (3.50) may be tested through the computed t ratio for $\widehat{\alpha}_2^*$ in

$$(1 + L^2) y_t^* = \alpha_2^* y_{t-2}^* + v_t. \tag{3.53}$$

The null hypothesis is $\alpha_2^* = 0$ with the alternative of stationarity implying $\alpha_2^* > 0$. With the use of the double subscript notation as in (3.51), then under the null hypothesis,

$$t(\widehat{\alpha}_2^*) = \left(\sum_{s=1}^{2} \sum_{\tau=1}^{T_\tau} v_{s\tau} y_{s,\tau-1}^* \right) \Big/ \left\{ \widetilde{\sigma} \left[\sum_{s=1}^{2} \sum_{\tau=1}^{T_\tau} (y_{s,\tau-1}^*)^2 \right]^{1/2} \right\}. \tag{3.54}$$

Applying the asymptotic distributional results discussed in the last subsection to this case implies, after scaling numerator and denominator of (3.54) by T_τ^{-1}, that

$$t(\widehat{\alpha}_2^*) \Rightarrow -\left[\sum_{s=1}^{2} \int_0^1 W_s(r)\, dW_s(r) \right] \Big/ \left\{ \left[\sum_{s=1}^{2} \int_0^1 [W_s(r)]^2\, dr \right]^{1/2} \right\}, \tag{3.55}$$

where $W_s(r)$ for $s = 1, 2$ are again independent standard Brownian motion processes. One important practical consequence of (3.55) is that with a simple change of sign, the DHF tables with $S = 2$ apply to the case of testing $\alpha_2^* = 0$ in (3.53). This is a straightforward implication of the "mirror image" property of a process with unit root of -1 compared with the usual unit root of 1, combined with arguments made above in our discussion of the DHF test.

Under the assumed DGP (3.50), we could also consider testing the null hypothesis relating to the omitted one-period lag, namely $\alpha_1^* = 0$, against the alternative $\alpha_1^* \neq 0$ through the test regression

$$(1 + L^2) y_t^* = \alpha_1^* y_{t-1}^* + v_t. \tag{3.56}$$

The test here is not strictly a unit root test, since the coefficient of unity on L^2 in (3.56) implies that the process contains two roots of modulus one, irrespective of the value of α_1^*. Now, denote the roots of quadratic $1 - \alpha_1^* L + L^2$ as $\kappa_1 \pm \kappa_2 i$ for the case of complex roots. The value of α_1^* affects κ_1 and κ_2, but not the modulus of the pair of roots since $\kappa_1^2 + \kappa_2^2 = 1$. The values of κ_1 and κ_2 do, however, lead to the spectral frequency associated with the complex unit root process, with $\kappa_1 = 0, \kappa_2 = 1$ yielding the roots $\pm i$ associated with frequency $\pi/2$. Thus, the test of $\alpha_1^* = 0$ is a test of the null hypothesis that the unit root process occurs at spectral frequency $\pi/2$ and hence that it contains a full cycle every four periods. The alternative hypothesis is two sided since there is no a priori information about the periodicity of the process under the alternative.

For the test regression (3.56),

$$
t(\widehat{\alpha_1^*}) = \frac{\sum_{t=1}^{T} v_t y_{t-1}^*}{\widetilde{\sigma} \left[\sum_{t=1}^{T} (y_{t-1}^*)^2 \right]^{1/2}} = \frac{\sum_{\tau=1}^{T_\tau} [v_{1\tau} y_{2,\tau-1}^* + v_{2\tau} y_{1\tau}^*]}{\widetilde{\sigma} \left\{ \sum_{\tau=1}^{T_\tau} [(y_{2,\tau-1}^*)^2 + (y_{1\tau}^*)^2] \right\}^{1/2}}
$$

$$
= \frac{T_\tau^{-1} \sum_{\tau=1}^{T_\tau} v_{2\tau} v_{1\tau} + T_\tau^{-1} \sum_{\tau=1}^{T_\tau} [v_{1\tau} y_{2,\tau-1}^* - v_{2\tau} y_{1,\tau-1}^*]}{\widetilde{\sigma} \left\{ T_\tau^{-2} \sum_{\tau=1}^{T_\tau} v_{1\tau}^2 + T_\tau^{-2} \sum_{\tau=1}^{T_\tau} [(y_{2,\tau-1}^*)^2 + (y_{1,\tau-1}^*)^2] \right\}^{1/2}}.
\tag{3.57}
$$

The second line here is obtained by dividing numerator and denominator by T_τ^{-1} and also using $y_{1\tau}^* = -y_{1,\tau-1}^* + v_{1\tau}$. This substitution for $y_{1\tau}^*$ is undertaken for essentially technical reasons, namely so that the terms are written in a form that ultimately allows the asymptotic distributions of (3.33) and (3.34) to be used. Also note that $T_\tau^{-1} \sum_{\tau=1}^{T_\tau} v_{2\tau} v_{1\tau}$ and $T_\tau^{-2} \sum_{\tau=1}^{T_\tau} v_{1\tau}^2$ can be neglected asymptotically. The former is asymptotically zero as it converges to the (zero) covariance between the independent processes $v_{1\tau}, v_{2\tau}$. For the latter, note that $T_\tau^{-1} \sum_{\tau=1}^{T_\tau} v_{1\tau}^2$ converges to σ^2 and hence this quantity divided by T_τ converges to zero.

For the sum $\sum_{\tau=1}^{T_\tau} v_{s\tau} y_{q,\tau-1}^*$, we can use the same approach as in the preceding subsection. Thus, with $\varepsilon_{s\tau} = (-1)^\tau v_{s\tau}$, it follows that

$$\sum_{\tau=1}^{T_\tau} v_{s\tau} y_{q,\tau-1}^* = \sum_{\tau=1}^{T_\tau} (-1)^\tau \varepsilon_{s\tau} \sum_{j=0}^{\tau-2} (-1)^j (-1)^{\tau-j-1} \varepsilon_{q,\tau-j-1}$$

$$= -\sum_{\tau=1}^{T_\tau} \varepsilon_{s\tau} y_{q,\tau-1},$$

where $y_{q\tau} = \sum_{j=1}^{\tau} \varepsilon_{qj}$. Notice, however, that the negative sign is now irrelevant in terms of the distribution because $\varepsilon_{s\tau}$ and $y_{q,\tau-1}$ are independent processes ($s \neq q$), with $\varepsilon_{s\tau}$ and $-\varepsilon_{s\tau}$ having identical distributions. Using (3.33), we can conclude that, for the processes of (3.51),

$$T_\tau^{-1} \sum_{\tau=1}^{T_\tau} v_{s\tau} y_{q,\tau-1}^* \Rightarrow \sigma^2 \int_0^1 W_q(r)\, dW_s(r), \qquad q, s = 1, 2, \quad (3.58)$$

and hence from (3.57),

$$t(\widehat{\alpha}_1^*) \Rightarrow \frac{\int_0^1 W_2(r)\, dW_1(r) - \int_0^1 W_1(r)\, dW_2(r)}{\left\{ \sum_{s=1}^2 \int_0^1 [W_s(r)]^2\, dr \right\}^{1/2}}. \quad (3.59)$$

Since $W_1(r)$ and $W_2(r)$ are identically distributed, it follows that $t(\widehat{\alpha}_1^*)$ is symmetrically distributed around zero.

Indeed, the results in (3.59) and (3.55) continue to apply for testing $\alpha_1^* = 0$ or $\alpha_2^* = 0$ (respectively) in the combined test regression

$$(1 + L^2)y_t^* = \alpha_1^* y_{t-1}^* + \alpha_2^* y_{t-2}^* + v_t, \quad (3.60)$$

because the regressors y_{t-1}^* and y_{t-2}^* are asymptotically orthogonal. This orthogonality is analogous to that between the HEGY variables $y_t^{(3)}$ and $y_t^{(4)} = -y_{t-1}^{(3)}$ noted in Subsection 3.2.4. Again for readers interested in the details, note that the scaled matrix of regressors $T^{-2} X' X$ for the regression (3.60) has the single off-diagonal term

$$\frac{1}{T^2} \sum_{t=1}^T y_{t-1}^* y_{t-2}^* = \frac{4}{T_\tau^2} \sum_{\tau=1}^{T_\tau} [y_{2,\tau-1}^* y_{1,\tau-1}^* + y_{1\tau}^* y_{2,\tau-1}^*],$$

$$= \frac{4}{T_\tau^2} \sum_{\tau=1}^{T_\tau} [y_{2,\tau-1}^* y_{1,\tau-1}^* - y_{1,\tau-1}^* y_{2,\tau-1}^* + v_{1\tau} y_{2,\tau-1}^*],$$

$$= \frac{4}{T_\tau^2} \sum_{\tau=1}^{T_\tau} v_{1\tau} y_{2,\tau-1}^*.$$

To obtain this result, we use $T = 2T_\tau$, and, after writing in our double subscript notation, the equation $y_{1\tau}^* = -y_{1,\tau-1}^* + v_{1\tau}$ from (3.51) is used. Although the distributional result of (3.58) states that $T_\tau^{-1} \sum_{t=1}^{T_\tau} v_{1\tau} y_{2,\tau-1}^* \Rightarrow \sigma^2 \int_0^1 W_2(r) dW_1(r)$, the final expression of interest to us involves scaling by T_τ^{-2} rather than T_τ^{-1}. Hence, $T^{-2} \sum_{t=1}^T y_{t-1}^* y_{t-2}^*$ converges asymptotically to its expected value of zero.

It is worth emphasizing that the asymptotic orthogonality of the regressors in (3.60) is a powerful property that substantially simplifies the analysis. As discussed in Subsection 3.2.5, this property does not arise when independent random walk processes are considered.

Before leaving this discussion of complex unit roots, an introduction to a different approach will later prove useful. This approach explicitly considers the complex unit roots through

$$(1 + L^2) y_t^* = (1 + iL)(1 - iL) y_t^* = v_t.$$

Now, if we impose the factor $(1 + iL)$ through the construction of the complex variable $x_t^{(1)} = (1 + iL) y_t^* = y_t^* + i y_{t-1}^*$, then we might consider testing the other complex factor through the regression

$$(1 - iL) x_t^{(1)} = (1 + L^2) y_t^* = (\pi_R + i\pi_I) x_{t-1}^{(1)} + v_t, \tag{3.61}$$

where the right-hand side of (3.61) allows the coefficient of the complex variable $x_{t-1}^{(1)}$ to have both a real part and an imaginary part. Under the unit root null hypothesis of interest, the right-hand side is simply v_t, so the formal null hypothesis is $\pi_R = \pi_I = 0$. Under the alternative hypothesis of stationarity, $\pi_I < 0$ (required for stationarity of the complex root), but π_R could be either positive or negative.

The complex root of i that is tested through (3.61) has a complex conjugate pair, namely the root $-i$. Therefore, corresponding to this equation is

$$(1 + iL) x_t^{(2)} = (1 + L^2) y_t^* = (\pi_R - i\pi_I) x_{t-1}^{(2)} + v_t, \tag{3.62}$$

where we define $x_t^{(2)} = (1 - iL) y_t^* = y_t^* - i y_{t-1}^*$. Intentionally, the coefficients on the right-hand of (3.61) and (3.62) are a complex conjugate pair; this must be the case because the regressor variables, namely $x_{t-1}^{(1)}$ and $x_{t-1}^{(2)}$, are themselves a pair of complex variables while the dependent variable $(1 + L^2) y_t^*$ is the same in both cases. That is, (3.61) and (3.62) are equivalent.

Combining the last two equations in a single test test regression leads to

$$\begin{aligned} (1 + L^2) y_t^* &= (\pi_R + i\pi_I) x_{t-1}^{(1)} + (\pi_R - i\pi_I) x_{t-1}^{(2)} + v_t, \\ &= 2\pi_R y_{t-1}^* + 2\pi_I y_{t-2}^* + v_t, \end{aligned} \tag{3.63}$$

when we replace $x_{t-1}^{(1)}$ and $x_{t-1}^{(2)}$ by their definitions in terms of y_{t-1}^* and y_{t-2}^*. It is immediately apparent that (3.63) is equivalent to (3.60). The interpretation of the latter now is that the coefficient of y_{t-2}^* tests the hypothesis that the

coefficient of the imaginary part of the complex root is, indeed, unity, while the
coefficient of y_{t-1}^* tests the hypothesis that this root contains no real component.

3.3.4   HEGY Test

As discussed in Subsection 3.2.4, the HEGY approach of Hylleberg et al. (1990)
is based on the factorization (3.17) for quarterly data. In this quarterly case, their
test regression has the form[1]

$$\Delta_4 y_t = \pi_1 y_{t-1}^{(1)} - \pi_2 y_{t-1}^{(2)} - \pi_3 y_{t-2}^{(3)} - \pi_4 y_{t-1}^{(3)} + \varepsilon_t, \qquad t = 1, 2, \ldots, T,$$
$$(3.64)$$

where $y_t^{(1)}$, $y_t^{(2)}$, and $y_t^{(3)}$ are defined in (3.16). The overall null hypothesis
$y_t \sim SI(1)$ implies $\pi_1 = \pi_2 = \pi_3 = \pi_4 = 0$ and hence $\Delta_4 y_t = \varepsilon_t$ as for the
DHF test with $S = 4$.

The form of (3.64) can be motivated by considering the three factors of
$\Delta_4 = (1 - L)(1 + L)(1 + L^2)$ one by one. Imposing the seasonal unit roots -1
and $\pm i$, through the variable $y_t^{(1)}$, we might consider a test of $\phi_1 = 1$ against
$\phi_1 < 1$ in

$$(1 - \phi_1 L)y_t^{(1)} = \varepsilon_t.$$

Using the definition $\pi_1 = \phi_1 - 1$, and noting that $(1 - L)y_t^{(1)} = \Delta_4 y_t$, leads to
a regression of the Dickey–Fuller type, namely

$$\Delta_4 y_t = \pi_1 y_{t-1}^{(1)} + \varepsilon_t,$$
$$(3.65)$$

where we test $\pi_1 = 0$ against $\pi_1 < 0$. Similarly, imposing the roots 1 and $\pm i$
through the variable $y_t^{(2)}$, testing the root -1 would lead to a test of $\phi_2 = 1$
against $\phi_2 < 1$ in

$$(1 + \phi_2 L)y_t^{(2)} = \varepsilon_t.$$

Defining $\pi_2 = \phi_2 - 1$ yields the form

$$\Delta_4 y_t = -\pi_2 y_{t-1}^{(2)} + \varepsilon_t,$$
$$(3.66)$$

when it is recognized that $(1 + L)y_t^{(2)} = \Delta_4 y_t$. The null and alternative hypotheses are $\pi_2 = 0$ and $\pi_2 < 0$, respectively. Notice that, if written as
$(1 + L)y_t^{(2)} = -\pi_2 y_{t-1}^{(2)} + \varepsilon_t$, then (3.66) has identical form to (3.45) for a test
of the unit root -1. This is by design, and, if we recall the discussion at the
end of Subsection 3.3.2, the advantage of testing π_2 in (3.66) is that the usual
Dickey–Fuller tables apply immediately.

[1] The test regression defined by HEGY differs slightly from (3.64) in that the minus sign is
incorporated into the definition of $y_t^{(2)}$ and $y_t^{(3)}$. This clearly affects only the signs of the
corresponding coefficients.

Finally, imposing the roots ± 1 through $y_t^{(3)} = y_t - y_{t-2}$, we wish to test $\phi_3 = 1$ and $\phi_4 = 0$ in

$$(1 + \phi_4 L + \phi_3 L^2) y_t^{(3)} = \varepsilon_t,$$

and, with $\pi_3 = \phi_3 - 1$ and $\pi_4 = \phi_4$, we obtain

$$\Delta_4 y_t = -\pi_3 y_{t-2}^{(3)} - \pi_4 y_{t-1}^{(3)} + \varepsilon_t. \tag{3.67}$$

Testing against stationarity implies null and alternative hypotheses of $\pi_3 = 0$ and $\pi_3 < 0$. However, while $\pi_4 = 0$ is also indicated under the null hypothesis, the alternative here is $\pi_4 \neq 0$. The reasoning for this two-sided alternative is precisely the same as that for the lag 1 coefficient α_1^* in the test regression of (3.56). It should be noted that, since $\Delta_4 y_t = (1 + L^2) y_t^{(3)}$, the test regression (3.67) is identical (when account is taken of signs) to the combined regression (3.60) discussed for the complex roots case in Subsection 3.3.3. Once again, the DHF tables for $S = 2$ apply without modification for a test of π_3.

Because of the asymptotic orthogonality of the regressors in (3.65), (3.66), and (3.67), these can be combined into the single test regression (3.64) without any effect on the asymptotic properties of the coefficient estimators. This asymptotic orthogonality for the HEGY regressors was noted at the end of Subsection 3.2.4.

To summarize, and building on our discussions of Subsections 3.3.2 and 3.3.3, we state that under the null hypothesis $\pi_1 = 0$, $y_t^{(1)}$ in (3.65) is a conventional random walk process and hence the asymptotic distribution of (3.27) applies. We denote the corresponding standard Brownian motion in this case as $W_1^*(r)$, so (3.27) becomes $T^{-1/2} y_t^{(1)} \Rightarrow \sigma W_1^*(r)$. Therefore, a test of the hypothesis $\pi_1 = 0$ in (3.65) is a Dickey–Fuller test with

$$t(\widehat{\pi}_1) \Rightarrow \left[\int_0^1 W_1^*(r) \, dW_1^*(r) \right] \Big/ \left\{ \int_0^1 [W_1^*(r)]^2 \, dr \right\}^{1/2}. \tag{3.68}$$

This is the conventional Dickey–Fuller distribution, tabulated by Fuller (1996).

Under the null hypothesis $\pi_2 = 0$, (3.66) implies

$$t(\widehat{\pi}_2) = -\left[\sum_{t=1}^{T} \varepsilon_t y_{t-1}^{(2)} \right] \Big/ \left\{ \widetilde{\sigma} \left[\sum_{t=1}^{T} \left(y_{t-1}^{(2)} \right)^2 \right]^{1/2} \right\}. \tag{3.69}$$

Our discussion of testing the unit root of -1, and particularly (3.49), immediately yields

$$t(\widehat{\pi}_2) \Rightarrow \left[\int_0^1 W_2^*(r) \, dW_2^*(r) \right] \Big/ \left\{ \int_0^1 [W_2^*(r)]^2 \, dr \right\}^{1/2}. \tag{3.70}$$

As already remarked, this is the conventional Dickey–Fuller distribution tabulated by Fuller (1996), with the minus sign of (3.49) being taken into account by use of the regressor $-y_{t-1}^{(2)}$ in (3.66). Similarly, for testing the complex pair of unit roots through (3.67), we can use the results of Subsection 3.3.3. Noting again the minus signs of (3.67), we have

$$t(\widehat{\pi}_3) \Rightarrow \frac{\int_0^1 W_3^*(r)\,dW_3^*(r) + \int_0^1 W_4^*(r)\,dW_4^*(r)}{\left\{\int_0^1 [W_3^*(r)]^2 dr + \int_0^1 [W_4^*(r)]^2 dr\right\}^{1/2}}, \tag{3.71}$$

by applying (3.55). Similarly, from (3.59) with a change of sign,

$$t(\widehat{\pi}_4) \Rightarrow \frac{\int_0^1 W_3^*(r)\,dW_4^*(r) - \int_0^1 W_4^*(r)\,dW_3^*(r)}{\left\{\int_0^1 [W_3^*(r)]^2 dr + \int_0^1 [W_4^*(r)]^2 dr\right\}^{1/2}}. \tag{3.72}$$

Notice also that the relevant standard Brownian motions in these cases are denoted by $W_2^*(r)$, $W_3^*(r)$, and $W_4^*(r)$. Thus, $W_2^*(r)$ is associated with $y_t^{(2)}$, while $W_3^*(r)$ and $W_4^*(r)$ are associated with the two processes implicit in the complex unit root behavior of $y_t^{(3)}$.

The notation $W_i^*(r)$, $i = 1, \ldots, 4$ is used in the equations of this section to distinguish these from the Brownian motions $W_s(r)$, $s = 1, \ldots, 4$ associated with the sth quarter of each year in Chapter 2 and earlier sections of this chapter. The processes $W_i^*(r)$ are, however, linear combinations of the $W_s(r)$, with these linear combinations reflecting the definitional relationships for the HEGY variables in (3.16). Precise details of the relationships between the two sets of Brownian motions can be found in Smith and Taylor (1998) or Osborn and Rodrigues (1998).

HEGY suggest that π_3 and π_4 might be jointly tested, since they are both associated with the pair of nonstationary complex roots $\pm i$. Such joint testing might be accomplished by computing $F(\widehat{\pi}_3 \cap \widehat{\pi}_4)$ as for a standard F test, although the distribution will not, of course, be the standard F distribution. Again because of the asymptotic independence of $\widehat{\pi}_3$ and $\widehat{\pi}_4$ arising from the orthogonality of the regressors, Engle et al. (1993) show that the limiting distribution of $F(\widehat{\pi}_3 \cap \widehat{\pi}_4)$ is identical to that of $1/2[t^2(\widehat{\pi}_3) + t^2(\widehat{\pi}_4)]$, where the two individual components are given in (3.71) and (3.72).

Ghysels, Lee, and Noh (1994), or GLN, extend the HEGY approach for quarterly data by proposing the joint test statistics $F(\widehat{\pi}_1 \cap \widehat{\pi}_2 \cap \widehat{\pi}_3 \cap \widehat{\pi}_4)$ and $F(\widehat{\pi}_2 \cap \widehat{\pi}_3 \cap \widehat{\pi}_4)$, the former being an overall test of the null hypothesis $y_t \sim SI(1)$ and the latter a joint test of the seasonal unit roots implied by the summation operator $1 + L + L^2 + L^3$. Once again because of the asymptotic orthogonality of the HEGY regressors, $F(\widehat{\pi}_1 \cap \widehat{\pi}_2 \cap \widehat{\pi}_3 \cap \widehat{\pi}_4)$ has the same asymptotic distribution as $1/4 \sum_{i=1}^4 [t(\widehat{\pi}_i)]^2$. Thus, from (3.68), (3.70), (3.71), and (3.72), it follows that (under the null hypothesis)

$$F(\widehat{\pi}_1 \cap \widehat{\pi}_2 \cap \widehat{\pi}_3 \cap \widehat{\pi}_4)$$

$$\Rightarrow \frac{1}{4} \left\{ \frac{\left[\int_0^1 W_1^*(r)\,dW_1^*(r) \right]^2}{\int_0^1 [W_1^*(r)]^2\,dr} + \frac{\left[\int_0^1 W_2^*(r)\,dW_2^*(r) \right]^2}{\int_0^1 [W_2^*(r)]^2\,dr} \right.$$

$$+ \frac{\left[\int_0^1 W_3^*(r)\,dW_3^*(r) + \int_0^1 W_4^*(r)\,dW_4^*(r) \right]^2}{\int_0^1 [W_3^*(r)]^2\,dr + \int_0^1 [W_4^*(r)]^2\,dr}$$

$$\left. + \frac{\left[\int_0^1 W_3^*(r)\,dW_4^*(r) - \int_0^1 W_4^*(r)\,dW_3^*(r) \right]^2}{\int_0^1 [W_3^*(r)]^2\,dr + \int_0^1 [W_4^*(r)]^2\,dr} \right\}. \tag{3.73}$$

Hence, with the use of the above discussion, $F(\widehat{\pi}_1 \cap \widehat{\pi}_2 \cap \widehat{\pi}_3 \cap \widehat{\pi}_4)$ is asymptotically distributed as the simple average of the squares of four distributions. Two of these are Dickey–Fuller distributions, arising from $t(\widehat{\pi}_1)$ and $t(\widehat{\pi}_2)$, one is a DHF distribution with $S = 2$, corresponding to $t(\widehat{\pi}_3)$, and the fourth is the symmetric distribution arising from $t(\widehat{\pi}_4)$. It is straightforward to see that a similar expression results for $F(\widehat{\pi}_2 \cap \widehat{\pi}_3 \cap \widehat{\pi}_4)$, which is a simple average of the squares of a Dickey–Fuller distribution, a DHF distribution with $S = 2$, and the symmetric distribution for $t(\widehat{\pi}_4)$.

GLN also observe that the usual test procedure of Dickey and Fuller, [or DF] (1979) can validly be applied in the presence of seasonal unit roots. There is, however, an important qualification, which is that this validity only holds if the regression contains sufficient augmentation to account for the seasonal unit roots. The essential reason derives from the factorization $1 - L^S = (1 - L)(1 + L + \cdots + L^{S-1})$. Therefore, the $SI(1)$ process $\Delta_S y_t = \varepsilon_t$ can be written as

$$\Delta_1 y_t = \alpha_1 y_{t-1} + \phi_1 \Delta_1 y_{t-1} + \cdots + \phi_{S-1} \Delta_1 y_{t-S+1} + \varepsilon_t, \tag{3.74}$$

with $\alpha_1 = 0$ and $\phi_1 = \cdots = \phi_{S-1} = -1$. With (3.74) applied as a unit root test regression, $t(\widehat{\alpha}_1)$ asymptotically follows the usual DF distribution. Nevertheless, $\phi_1, \ldots, \phi_{S-1}$ depend on the seasonal unit roots, so that the estimated values of these coefficients have nonstandard distributions. Rodrigues and Osborn (1999a) illustrate the adverse implications of failing to augment a Dickey–Fuller test regression with sufficient lags to account for seasonal unit roots.

Beaulieu and Miron (1993) and Franses (1991) develop the HEGY approach for the case of monthly data. As noted in Subsection 3.2.4, the transformed variables proposed by Beaulieu and Miron can be shown to be equivalent to the transformation of (3.15) applied to a sequence of twelve observations ending in period t and, thus, their variables share the asymptotic orthogonality of the HEGY definitions. This is not the case for the variables used by Franses, and hence the Beaulieu–Miron form is typically preferred. Beaulieu and Miron also present the asymptotic distributions for their estimated coefficients, noting that the t-type statistics corresponding to the two real roots of $+1$ and -1 each have the usual Dickey–Fuller form, while the remaining coefficients correspond to five pairs of complex roots. In this parameterization, each of the five pairs of complex roots leads to a $t(\widehat{\pi}_i)$ with a DHF distribution (again with $S = 2$) and a $t(\widehat{\pi}_i)$ with the distribution (3.72). Although both Beaulieu and Miron (1993) and Franses (1991) discuss the use of joint F-type statistics for the two coefficients corresponding to a pair of complex roots, neither considers the use of the F tests as in GLN to test the overall Δ_{12} filter or the eleven seasonal unit roots. These overall tests in the monthly context are considered by Taylor (1997), who supplies appropriate critical values. Another development of the HEGY approach is undertaken by Breitung and Franses (1998), who generalize HEGY tests to Phillips and Perron (1988) type statistics.

When considering overall F-type tests in the context of seasonal integration, one must bear in mind that such tests are two sided in nature, with the relevant alternative hypothesis in each case being that one or more of the unit root restrictions is not valid. Such invalidity of the null hypothesis could be due to a single root (or a complex pair of roots) being stationary or, indeed, greater than unity, with all other unit roots implied by Δ_S being valid. Thus, overall F tests of the $SI(1)$ null hypothesis cannot, strictly speaking, be interpreted as testing seasonal integration against the alternative hypothesis of stationarity for the process.

3.3.5   The Kunst Test

Kunst (1997) takes an apparently different approach to testing seasonal integration from that of HEGY. He uses the test regression

$$\Delta_S y_t = \phi_1 y_{t-1} + \cdots + \phi_{S-1} y_{t-S+1} + \alpha_S y_{t-S} + \varepsilon_t, \qquad t = 1, \ldots, T.$$
$$(3.75)$$

Clearly, this can be viewed as the DHF regression with the inclusion of the intermediate lags $y_{t-1}, \ldots, y_{t-S+1}$ that are assumed to have zero coefficients in (3.36). The null hypothesis of seasonal integration implies that $\phi_1 = \cdots = \phi_{S-1} = \alpha_S = 0$ in (3.75), and Kunst proposes that this hypothesis should be tested by using the joint statistic[2] $F(\widehat{\phi}_1 \cap \cdots \cap \widehat{\phi}_{S-1} \cap \widehat{\alpha}_S)$. As a unit root test, this does not have a conventional F distribution, but Kunst provides critical values for various values of S.

The asymptotics of Kunst's analysis are somewhat messy, and we again consider only the case $S = 4$. As discussed a number of times already, including

---

[2] In fact, Kunst does not compute his statistic by using the conventional F-statistic formula in that he chooses not to divide the numerator by S (which is n in his notation); see his equation (3.4).

in Subsection 3.2.1 above, under the seasonal integration null hypothesis, $\Delta_4 y_t = \varepsilon_t$ and y_{t-1}, y_{t-2}, y_{t-3}, y_{t-4} are generated from independent random walks. However, cross-product terms of the form $\sum_{t=1}^{T} y_{t-i} y_{t-j}$, $i < j$ consist of mixtures of cross products of these random walks $j - i$ quarters apart. For example, with the use of the two-subscript notation that indicates the month and year,

$$\sum_{t=1}^{T} y_{t-1} y_{t-2} = \sum_{\tau=1}^{T_\tau} (y_{4,\tau-1} y_{3,\tau-1} + y_{1\tau} y_{4,\tau-1} + y_{2\tau} y_{1\tau} + y_{3\tau} y_{2\tau}).$$

With the use of the asymptotic result (3.34) and, as $T = 4T_\tau$ so that $T^{-2} = T_\tau^{-2}/16$, therefore[3]

$$T^{-2} \sum_{t=1}^{T} y_{t-1} y_{t-2} \Rightarrow \frac{1}{16} \sigma^2 \int_0^1 [W_4(r)W_1(r) + W_2(r)W_1(r)$$

$$+ W_3(r)W_2(r) + W_4(r)W_3(r)]dr. \qquad (3.76)$$

Using the same logic, we can see that $T^{-2} \sum_{t=1}^{T} y_{t-2} y_{t-3}$, $T^{-2} \sum_{t=1}^{T} y_{t-3} y_{t-4}$, and $T^{-2} \sum_{t=1}^{T} y_{t-1} y_{t-4}$ all have the asymptotic distribution given by (3.76). Further, following similar steps, we find

$$T^{-2} \sum_{t=1}^{T} y_{t-1} y_{t-3} \Rightarrow \frac{1}{8} \sigma^2 \int_0^1 [W_1(r)W_3(r) + W_2(r)W_4(r)]dr.$$

$$(3.77)$$

The most important conclusion to be drawn from these last two equations is that the off-diagonal terms of the scaled cross-product matrix for the regressors, $T^{-2}(X'X)$, are nonzero for the Kunst regression. The reasoning is the same as that discussed at the end of Subsection 3.2.5. Consequently, the asymptotic orthogonality that simplified the analysis in the HEGY regression (3.64) does not apply for the Kunst regression (3.75). These results generalize to any S, so that the inclusion of the redundant lags $y_{t-1}, \ldots, y_{t-S+1}$ in (3.75) has a nontrivial asymptotic effect on the properties of the estimated coefficients. As a direct consequence of the nondiagonality of $T^{-2}(X'X)$, $t(\widehat{\alpha}_S)$ does not follow the DHF distribution of (3.40).

Nevertheless, as noted above, Kunst is primarily concerned with the distribution of the joint test statistic $F(\widehat{\phi}_1 \cap \cdots \cap \widehat{\phi}_{S-1} \cap \widehat{\alpha}_S)$. Although apparently overlooked by Kunst, it is easy to see that this joint test is identical to the joint test of all coefficients that arises in the HEGY framework. For the quarterly

---

[3] Notice that $\sum_{\tau=1}^{T_\tau} y_{1\tau} y_{4,\tau-1} = \sum_{\tau=1}^{T_\tau} \varepsilon_{1\tau} y_{4,\tau-1} + \sum_{\tau=1}^{T_\tau} y_{1,\tau-1} y_{4,\tau-1}$, but $T_\tau^{-2} \sum_{\tau=1}^{T_\tau} \varepsilon_{1\tau} y_{4,\tau-1} \Rightarrow 0$.

case, note, first, that the regressors $y_{t-1}^{(1)}$, $y_{t-1}^{(2)}$, $y_{t-2}^{(3)}$, and $y_{t-1}^{(3)}$ of (3.64) are defined in Subsection 3.2.4 as nonsingular linear transformations of y_{t-1}, y_{t-2}, y_{t-3}, and y_{t-4}. However, F-type statistics, which are based on residual sums of squares, are invariant to such nonsingular linear transformations of the regressors. Therefore, the residuals computed from (3.64) and from (3.75) with $S = 4$ must be identical, implying numerically identical values for the statistics $F(\widehat{\pi}_1 \cap \widehat{\pi}_2 \cap \widehat{\pi}_3 \cap \widehat{\pi}_4)$ and $F(\widehat{\phi}_1 \cap \widehat{\phi}_2 \cap \widehat{\phi}_3 \cap \widehat{\alpha}_4)$ applied to a given data set. Also, a comparison of the percentiles tabulated by Ghysels et al. (1994) for the distribution of the former and Kunst with $S = 4$ for the latter verifies that these are effectively identical.[4] Naturally, these results carry over to the monthly case, with the HEGY F statistic of Taylor (1997) being equivalent to that of Kunst with $S = 12$. Kunst does, however, provide critical values for additional cases, including $S = 7$, which is relevant for testing the null hypothesis that a daily series is seasonally integrated at a period of 1 week.

3.3.6   The Osborn–Chui–Smith–Birchenhall Test

Although originally proposed for testing higher orders of seasonal integration, the test of Osborn et al. (1988), often referred to as OCSB, can be implemented as a test of first-order seasonal integration. To see this, note that (in terms of general S), OCSB discuss the test regression

$$\Delta_1 \Delta_S y_t = \beta_1 \Delta_S y_{t-1} + \beta_2 \Delta_1 y_{t-s} + \varepsilon_t, \qquad t = 1, \ldots, T, \quad (3.78)$$

which had earlier been considered by Hasza and Fuller (1982). Using the identity $\Delta_S = \Delta_1(1 + L + \cdots + L^{S-1})$, Rodrigues and Osborn (1999a) note that $\Delta_S y_{t-1} = (1 + L + \cdots + L^{S-1})\Delta_1 y_{t-1}$ and hence all variables in (3.78) impose first differences. An interpretation of the OCSB test, therefore, is that it tests seasonal integration in a process that has been first differenced. Although the OCSB test will be discussed in Subsection 3.4.5 in the context of its original purpose, here we consider its application directly to the observed data. That is, following Rodrigues and Osborn, we examine the test regression

$$\Delta_S y_t = \beta_1(1 + L + \cdots + L^{S-1})y_{t-1} + \beta_2 y_{t-s} + \varepsilon_t, \qquad t = 1, \ldots, T,$$
$$(3.79)$$

which is proposed by Rodrigues and Osborn as a vehicle for testing the null hypothesis $y_t \sim SI(1)$. Unlike all the other test regressions considered so far,

---

[4] Once again, because of the definition of his F-type statistic, the Kunst (1997) percentiles have to be divided by four to be comparable with those of Ghysels et al. (1994). In the monthly case, the Kunst values have to be divided by twelve for comparison with Taylor (1997). Since these percentiles are obtained from Monte Carlo simulations, they will not be identical across different studies.

the regressors of (3.79) are not independent because the first regressor $(y_{t-1} + \cdots + y_{t-S})$ includes the second regressor y_{t-S}.

In the context of (3.79), the test statistic $t(\widehat{\beta}_1)$ can be used to test the validity of the zero frequency unit root. That is, it can be viewed as a Dickey–Fuller test applied to the annual sum series $(1 + L + \cdots + L^{S-1})y_t$. If a test against stationarity is desired, then the appropriate alternative to $\beta_1 = 0$ is $\beta_1 < 0$. Similarly, $t(\widehat{\beta}_2)$ can be interpreted as a DHF test of $\beta_2 = 0$ against the alternative $\beta_2 < 0$. Such interpretations must be made with care, however, because (as we shall see) the Dickey–Fuller and DHF critical values are not valid in the context of this test regression. Indeed, OCSB incorrectly give the impression that these critical values remain valid.

For simplicity, we consider $S = 4$. A fuller and more general analysis of the asymptotic distributions for the OCSB test is contained in Osborn and Rodrigues (1998). In the quarterly case, the first diagonal term of the regressor cross-product matrix $X'X$ for this test is $\sum_{t=1}^{T}(y_{t-1} + \cdots + y_{t-4})^2 = \sum_{t=1}^{T}(y_{t-1}^{(1)})^2$, where $y_t^{(1)}$ is defined by (3.16) and is the first regressor of the HEGY regression (3.64). As discussed in the context of the HEGY test, $y_t^{(1)}$ follows a random walk under the $SI(1)$ null hypothesis and

$$T^{-2}\sum_{t=1}^{T}\left[y_{t-1}^{(1)}\right]^2 \Rightarrow \sigma^2 \int_0^1 [W_1^*(r)]^2\, dr.$$

The second diagonal element of the scaled regressor cross-product matrix in this regression is

$$T^{-2}\sum_{t=1}^{T}(y_{t-4})^2 = (4T_\tau)^{-2}\sum_{\tau=1}^{T_\tau}\sum_{s=1}^{4}(y_{s,\tau-1})^2$$

$$\Rightarrow \frac{\sigma^2}{16}\sum_{s=1}^{4}\int_0^1 [W_s(r)]^2\, dr,$$

where this result uses the fact that each $y_{s\tau}$, $s = 1, \ldots, 4$ is a random walk process. Unfortunately, however, the off-diagonal term of $T^{-2}X'X$ is not asymptotically zero and, in the quarterly case, is

$$T^{-2}\sum_{t=1}^{T}y_{t-4}(y_{t-1} + \cdots + y_{t-4}) = T^{-2}\left[\sum_{t=1}^{T}(y_{t-4})^2 + \sum_{i=1}^{3}\sum_{t=1}^{T}y_{t-i}y_{t-4}\right].$$

Not only is this nonzero because of the term $T^{-2}\sum_{t=1}^{T}(y_{t-4})^2$ (which results from y_{t-4} being part of the first regressor), but also because the terms $T^{-2}\sum_{t=1}^{T}y_{t-i}y_{t-4}$ ($i = 1, 2, 3$) converge to the random variables (3.76) or (3.77) as discussed in Subsection 3.3.5. While expressions can be obtained for the asymptotic distributions of $t(\widehat{\beta}_1)$ and $t(\widehat{\beta}_2)$, these are not especially illuminating, and we will not consider them here. The important point is that the DF and DHF distributions would apply for $t(\widehat{\beta}_1)$ and $t(\widehat{\beta}_2)$, respectively, if $T^{-2}X'X$ is asymptotically diagonal, but this diagonality property does not hold. Therefore, specific tables of critical values are required for these OCSB tests.

The OCSB test can be applied as a joint test of $\beta_1 = \beta_2 = 0$. In the context of (3.78), this is one form of the test originally proposed by Hasza and Fuller (1982). Here we continue to consider the form (3.79) and note that this regression can be reparameterized as

$$\Delta_S y_t = \beta^*(y_{t-1} + \cdots + y_{t-s+1}) + \alpha_S y_{t-s} + \varepsilon_t, \qquad t = 1, \ldots, T.$$
(3.80)

Since the regressors here are obtained as a nonsingular linear transformation of those in (3.79), the joint test statistics $F(\widehat{\beta}_1 \cap \widehat{\beta}_2)$ and $F(\widehat{\beta}^* \cap \widehat{\alpha}_s)$ will be numerically identical. Here (3.80) can, however, be viewed as a restricted version of Kunst's test regression (3.75), with the single coefficient applied to $y_{t-1} + \cdots + y_{t-s+1}$ implying the restriction $\phi_1 = \cdots = \phi_{S-1}$. Although the validity of the restriction may need to be verified, the use of a single parameter rather than the separate coefficients $\phi_1, \ldots, \phi_{S-1}$ may lead to increased power over the joint test of Kunst or the equivalent joint test in the HEGY framework discussed in Subsection 3.3.4. This is considered again later in this chapter, namely in the discussion of Monte Carlo results in Section 3.5.

3.3.7 Multiple Tests and Levels of Significance

It is notable that many of the seasonal unit root tests discussed above involve tests on more than one coefficient. For example, for the HEGY test regression (3.64), Hylleberg et al. (1990) recommend that one-sided tests should be applied for π_1 and π_2, with (π_3, π_4) either tested sequentially or jointly. The rationale for using one-sided tests for π_1, π_2, and π_3 is that each of these roots is tested against stationarity, which is not the case when a joint F type is applied. The overall seasonal integration null hypothesis is then rejected in favor of stationarity for the process only if the null hypothesis is rejected for each of these three tests. Many applied researchers have followed HEGY's advice, apparently failing to recognize the implications of this strategy for the overall level of significance for the implied joint test of $\pi_1 = \pi_2 = \pi_3 = \pi_4 = 0$.

Let us assume that separate tests are applied to π_1 and π_2, with a joint test applied to (π_3, π_4), with each of these three tests applied at some level of significance α. Conveniently, these tests are mutually independent, as a result of the asymptotic orthogonality of the regressors. Therefore, the overall probability of not rejecting the $SI(1)$ null hypothesis when it is true is $(1 - \alpha)^3$, which is approximately $1 - 3\alpha$ for α small. Thus, with the conventional $\alpha = 0.05$, the implied level of significance for the overall test is $1 - 0.95^3 = 0.14$. With monthly data, the issue is even more important. If the level of significance α is used with separate tests applied to π_1, π_2, and each of the pairs (π_i, π_{i+1}) ($i = 3, 5, 7, 9, 11$), then the implied overall test of the $SI(1)$ null hypothesis has significance level $1 - (1 - \alpha)^7$, or approximately 7α. In this case, $\alpha = 0.05$ implies an overall level of significance of 0.30. Although Dickey (1993) draws attention to the implication of using multiple tests in the HEGY framework, many applied studies have ignored this.

If an overall level of α is desired, then a simple way to (approximately) yield this would be to use a level of α/n for each individual test, where n is the number of independent tests applied (the approach of jointly testing each pair of complex unit roots and testing each real root separately yields $n = 3$ in the quarterly case and $n = 7$ with monthly data). For this to be made operational, additional critical values have to be tabulated for the distributions of $t(\widehat{\pi}_1)$ and $F(\widehat{\pi}_3 \cap \widehat{\pi}_4)$ for levels of significance other than the conventional 1% and 5%.

Unfortunately, this relatively simple analysis cannot be applied to other cases, since the asymptotic orthogonality that applies to the HEGY regressors does not then generally apply. However, if it is desired to conduct separate tests in order to use one-sided alternatives, then progress can be made through the use of the Bonferroni inequality. This states that, when n separate tests are applied, with levels of significance of α_i ($i = 1, \ldots, n$), then the overall level of significance for the implied joint test is at most $\sum_{i=1}^{n} \alpha_i$. This sets an upper bound to the implied level of significance and does not depend on the independence of the individual tests. Savin (1984) provides a good discussion of the issues involved in multiple hypothesis tests in the context of a linear regression model.

Application of the Bonferroni inequality could be made to set a bound for the overall level of significance when separate tests are applied in the context of the Kunst or OCSB regressions. For the former, a preliminary joint test of $\phi_1 = \cdots = \phi_{S-1} = 0$ could be applied in (3.75), with a subsequent one-sided test computed by using $t(\widehat{\alpha}_S)$. Clearly, different levels of significance could be used for these two tests, to reflect the relative importance given to the preliminary test of intermediate lags and the DHF-type seasonal unit root test $t(\widehat{\alpha}_S)$. It should be noted, however, that both of these distributions are nonstandard. Further, as mentioned above, the latter does not follow the asymptotic DHF distribution because of the presence of the lags $y_{t-1}, \ldots, y_{t-S+1}$ in the test regression. When the same logic is applied to the OCSB regression (3.79), separate one-sided tests could be performed by using $t(\widehat{\beta}_1)$ and $t(\widehat{\beta}_2)$, while also controlling the overall level of significance.

Of course, joint tests can be performed directly on the parameters of the relevant test regressions through the use of F statistics to control the overall level of significance. However, this approach not only fails to recognize the (possibly) one-sided nature of the alternative hypothesis for a specific coefficient, but it also implicitly treats the test on each coefficient with equal importance. Neither of these may be desirable when testing the overall seasonal unit root null hypothesis.

The general message is that one must bear in mind the impact of multiple tests when applying seasonal unit root tests. To date, however, these issues have received relatively little attention in this literature.

3.4   Extensions

Building on the material in Section 3.3, here we consider important extensions in order to examine a number of issues that are likely to arise in practice. These issues include the role of deterministic components in seasonal unit root test regressions, the effect of structural breaks in these deterministic components, and testing for higher orders of nonstationarity. First, however, we look very briefly at the issue of additional dynamics from a theoretical perspective.

3.4.1   Additional Dynamics

To keep the presentation as simple as possible, the unit root test regressions of Section 3.3 include no dynamic terms beyond those essential to the specific test. Typically, however, additional dynamics will be included to account for the autocorrelation properties of the dependent variable. Provided that these additional dynamics comprise only stationary variables, it is generally assumed that the asymptotic properties of the unit root tests are unaffected. Therefore, lags of $\Delta_s y_t$ are usually included as additional regressors in the test regressions considered in Section 3.3. Similarly, for testing the null hypothesis $y_t \sim I(2, 1)$ as below in Subsection 3.4.5, lags of $\Delta_1 \Delta_s y_t$ are included.

In the context of the quarterly HEGY test regression, Burridge and Taylor (2000) provide a theoretical and empirical analysis of the role of additional dynamics and find that the consequences may be serious. In particular, they establish that the asymptotic distributions for the t-type statistics associated with the complex pair of unit roots are not invariant to disturbance autocorrelation and lag augmentation does not remove the problem. Joint F-type statistics which consider these roots as a pair are, however, invariant. This provides a strong argument, in addition to those of Subsection 3.3.4, for the use of joint tests in the HEGY framework.

A further important practical question is how the order of autoregressive augmentation should be specified, but there appears to have been relatively little discussion of this issue in the context of seasonal processes. In a nonseasonal context, Hall (1994) examines the conditions under which the augmented

Dickey–Fuller unit root test retains its usual asymptotic distribution when the order of augmentation is based on the data. Reassuringly, the theoretical finding is that common procedures (including starting with a high order of possible augmentation and testing down by using conventional test statistics or specifying the lag order by using the Akaike or Schwarz criteria) imply that the asymptotic Dickey–Fuller test distribution continues to hold. The situation is, however, less clear in practice. These complications are discussed in Section 3.5 in the context of Monte Carlo studies.

3.4.2 Deterministic Components

The inclusion of deterministic components in the test regression has nontrivial effects on the asymptotic and finite sample properties of the Dickey–Fuller test statistic; this is discussed by, for example, Hamilton (1994a) or Banerjee et al. (1993). Analogous implications apply when deterministic terms are included in the seasonal unit root test regressions of Section 3.3. That is, even if the true DGP contains no deterministic components and has zero starting values, then the inclusion of a constant, seasonal dummy variables, or a linear trend will affect at least some of the asymptotic distributions presented above.

The inclusion of S seasonal dummy variables in any of the above test regressions has the effect of attaching an intercept to each season. In effect, each of the S random walks $y_{s\tau}$ ($s = 1, \ldots, S$) is then replaced by the demeaned random walk $\tilde{y}_{s\tau} = y_{s\tau} - \bar{y}_s$, where $\bar{y}_s$ is the sample mean of the T_τ observations for season s. From (3.31), the consequence is that the Brownian motion $W_s(r)$ must be replaced by demeaned Brownian motion $\tilde{W}_s(r) = W_s(r) - \int_0^1 W_s(r)dr$ when seasonal dummy variables are included in the test regressions of Section 3.3. The inclusion of a trend (in addition to the seasonal dummy variables) leads to the use of a demeaned and detrended Brownian motion, where the detrending involves all $W_s(r)$, $s = 1, \ldots, S$; see Smith and Taylor (1998).

The asymptotic distributions in the case of the HEGY test regression involve the Brownian motions $W_i^*(r)$, which (as noted in Subsection 3.3.4) are linear combinations of the $W_s(r)$, $s = 1, \ldots, S$. Since removing the mean from each series $y_{s\tau}$ ($s = 1, \ldots, S$) will remove the mean from the HEGY variables, then this implies that the asymptotic distributions for the HEGY tests will involve demeaned $\tilde{W}_i^*(r)$ when seasonal dummy variables are included in the test regression.

The situation is, however, different in relation to the trend. The removal of a common trend for all seasons affects only the zero frequency and hence only $W_1^*(r)$ is replaced by the corresponding demeaned and detrended Brownian motion. When the quarterly case is considered, the variables $y_t^{(2)}$ and $y_t^{(3)}$ are defined in terms of differences between the values $y_{s\tau}$ for different quarters, and this differencing will annihilate any common trend. Thus, the asymptotic distributions for the coefficients of these variables, namely $\widehat{\pi}_2$, $\widehat{\pi}_3$, and $\widehat{\pi}_4$, are not affected by the inclusion of a common trend. They are, however, affected if seasonally varying trends are included in the test regression.

A different (but complementary) perspective to deterministic components is given by considering similar tests, where a similar test is one for which the distribution of the statistic of interest is unaffected by nuisance parameters in the DGP. We argue in Subsection 3.2.1 above that a realistic seasonal random walk DGP may have nonzero starting values and a constant drift term. Under these circumstances, we would like to use similar tests that are invariant to the particular values taken by the S starting values and the drift. In a nonseasonal context, it has been recognized for some time that the introduction of a constant and trend is required to deliver the required invariance to the unit root test statistic [see, e.g., Nankervis and Savin (1985)]. Kiviet and Phillips (1992) examine this issue in a fairly general model. Applied to the seasonal context, the analysis of Kiviet and Phillips implies that seasonal dummy variables have to be included to yield invariance to the starting values, while a trend delivers invariance to the value of the (constant) drift. Further, the inclusion of these deterministic components allows the possibility under the alternative hypothesis that y_t is stationary around a (common) long-run linear trend plus deterministic seasonality.

Therefore, to test the $SI(1)$ null hypothesis, our general recommendation is that a linear trend and seasonal dummy variables should be included in the test regression. Seasonal trends are required only if the drift terms under the null hypothesis may be seasonally varying.[5] When the higher-order $I(2, 1)$ null hypothesis, discussed in Subsection 3.4.5 below, is tested, seasonal dummy variables alone will be sufficient to give invariance to starting values. A trend will be required in this case only if the $I(2, 1)$ process could contain a nonzero drift, which would then imply that $E(y_t)$ could include a quadratic trend.

3.4.3   *Structural Change in the Deterministic Components*

Since the seminal paper of Perron (1989), it has been recognized that a nonseasonal process that is stationary around a trend can appear to exhibit unit root properties if the deterministic component is subject to structural change that is not recognized. Recent analyses, including Ghysels (1994a), Smith and Otero (1997), and Franses and Vogelsang (1998), have extended these results to the seasonal case in which one or more of the seasonal dummy variable coefficients may exhibit structural change.

---

[5] See also Smith and Taylor (1999), who develop the use of seasonally varying trends in the HEGY context and propose the use of likelihood ratio seasonal unit root tests analogous to those of Dickey and Fuller (1981).

To allow for a single structural break, but assuming (for simplicity) no trend and i.i.d. disturbances, the seasonal unit root null hypothesis could be tested through the regression

$$\Delta_S y_t = \alpha_S y_{t-S} + \sum_{s=1}^{4} \gamma_s \delta_{st} + \sum_{s=1}^{4} \gamma_s^* \delta_{st}^* + \varepsilon_t, \tag{3.81}$$

where $\delta_{st}^* = 0$ for $t \leq T_B$ and $\delta_{st}^* = \delta_{st}$ for $t > T_B$. Thus, δ_{st}^* is a conventional seasonal dummy variable after time period T_B, but is zero up to this time. It is helpful to let $T_B = \lambda T$, so that λ is the proportion of the sample that occurs prior to the break. A test of the null hypothesis $\alpha_S = 0$ in (3.81) is then a DHF test (see Subsection 3.3.1) allowing for a structural break, with the alternative hypothesis permitting a change in the deterministic seasonal pattern after the break point T_B.

As noted by Ghysels (1994a) in this context, and by Perron (1989) in the nonseasonal case, the structural break permitted in (3.81) effectively allows (under the null hypothesis) a new seasonal random walk process to begin after time T_B. However, as mentioned on a number of occasions already, a seasonal random walk process consists of S separate unit root processes. Therefore, one set of S independent random walks applies until time T_B and another (independent) set of S random walks applies thereafter, so that the asymptotic distribution of the t ratio for $\widehat{\alpha}_S$ involves $2S$ independent Brownian motions. In particular, this t ratio asymptotically converges to

$$t(\widehat{\alpha}_S) \Rightarrow$$

$$\frac{\lambda \sum_{s=1}^{S} \int_0^1 \widetilde{W}_s(r) \, d\widetilde{W}_s(r) + (1-\lambda) \sum_{s=S+1}^{2S} \int_0^1 \widetilde{W}_s(r) \, d\widetilde{W}_s(r)}{\left(\lambda(1-\lambda)\{\lambda^2 \sum_{s=1}^{S} \int_0^1 [\widetilde{W}_s(r)]^2 \, dr + (1-\lambda)^2 \sum_{s=S+1}^{2S} \int_0^1 [\widetilde{W}_s(r)]^2 \, dr\}\right)^{1/2}},$$

$$\tag{3.82}$$

where $\widetilde{W}_s(r)$, $s = 1, 2, \ldots, 2S$ are independent demeaned Brownian motions. In the special case of $\lambda = 0.5$, (3.82) is a DHF distribution for $2S$ seasons with intercepts included in the test regression. This can be seen by a comparison of the expression that results when $\lambda = 0.5$ with (3.40), allowing for the role of seasonal dummies in the latter. Asymptotically, this distribution continues to apply when ε_t is replaced by a more general stochastic process z_t.

This analysis can be generalized to the case of other seasonal unit root tests, with the HEGY test for quarterly data being considered by Smith and Otero (1997) and by Franses and Vogelsang (1998). The study of Smith and Otero is a Monte Carlo one, whereas Franses and Vogelsang provide a theoretical discussion. It is, however, disappointing that these studies consider separately the test statistics $t(\widehat{\pi}_1)$, $t(\widehat{\pi}_2)$, and $F(\widehat{\pi}_3 \cap \widehat{\pi}_4)$, but not the overall statistic $F(\widehat{\pi}_1 \cap \widehat{\pi}_2 \cap \widehat{\pi}_3 \cap \widehat{\pi}_4)$. Following the approach of Zivot and Andrews (1992),

Franses and Vogelsang discuss how the break point T_B can be estimated from the data.

It is unsurprising that allowing for structural breaks affects the distributions of all HEGY test statistics, shifting the critical values for $t(\widehat{\pi}_1)$ and $t(\widehat{\pi}_2)$ to the left and that of $F(\widehat{\pi}_3 \cap \widehat{\pi}_4)$ to the right. As noted by Franses and Vogelsang, when the break point is estimated as that minimizing $t(\widehat{\pi}_1)$ or $t(\widehat{\pi}_2)$ as appropriate, their tabulated critical values appear to be empirically identical to those which apply to the unit root test in the nonseasonal case (Perron and Vogelsang, 1992). From the discussion of Subsection 3.3.4 about the relationship of the HEGY t ratios $t(\widehat{\pi}_1)$ and $t(\widehat{\pi}_2)$ with the conventional Dickey–Fuller unit root test statistic, this result holds no surprise.

Another perspective is provided by the analyical results of da Silva Lopes and Montanes (1999). They assume that the DGP is the deterministic season-ality model with structural breaks, but that the usual HEGY test is applied. It is interesting that all individual and joint HEGY test statistics are shown to di-verge asymptotically, with the single exception that $t(\widehat{\pi}_4) \Rightarrow 0$; recall that $t(\widehat{\pi}_4)$ is not a unit root test. However, the distributions are affected by the magnitude of the structural break and typically the test statistics are shifted toward zero by the occurrence of the break. The conclusions, therefore, are that asymptoti-cally the HEGY test statistics can distinguish between a structural break in the deterministic seasonal component and seasonal unit roots, but that empirically the presence of a structural break will reduce the power of these tests. Indeed, this analysis explains many of the findings in Smith and Otero (1997) related to power.

3.4.4   Near Seasonal Integration

Recent analyses in a nonseasonal context have enabled stationary and integrated processes to be seen as a continuum through the development of the theory of near integrated processes; see, in particular, Phillips (1987). A corresponding analysis has also been applied in a seasonal context.

For the seasonal AR(1) process

$$y_t = \phi_S y_{t-S} + \varepsilon_t,$$

near seasonal integration can be captured by defining

$$\phi_S = \exp(c/T_\tau) \simeq 1 + c/T_\tau,$$

where c is a constant noncentrality parameter and, as previously, $T_\tau = T/S$ is the number of observations available for each of the S seasons of the year. The case of $c = 0$ leads to a seasonally integrated process. Although $c < 0$ corresponds to stationarity, the process approaches seasonal integration as T_τ increases. With the use of this setup, the analyses of Chan (1989b), Perron

(1991), and Tanaka (1996) extend theory of near integration developed in the nonseasonal context. This enables the power properties of the DHF test to be examined in the context of a near seasonally integrated DGP.

Rodrigues (1999) takes this analysis further by considering the HEGY test. In this latter case, the asymptotic orthogonality of the HEGY regressors implies that different noncentrality parameters can be permitted, and hence the extent of nearness to nonstationarity can differ, at the zero and each of the seasonal frequencies. One application of this analysis is the derivation of confidence intervals for the individual zero frequency and seasonal roots of an observed seasonal process; Rodrigues and Osborn (1999b) present such confidence intervals for quarterly U.K. macroeconomic time series.

3.4.5 Higher Order Nonstationarity

Testing the $SI(1)$ null hypothesis has been discussed at some length, but it is also relevant to consider higher orders of nonstationarity. For example, the so-called airline model of Box and Jenkins (1976) has often been applied to describe seasonally unadjusted economic time series. This model assumes a DGP of the form

$$\Delta_1 \Delta_S y_t = (1 - \theta_1 L)(1 - \theta_S L^S)\varepsilon_t, \tag{3.83}$$

with $|\theta_1|$, $|\theta_S| < 1$. When higher-order nonstationarity may be present, it is appropriate to invoke a strategy of "testing down" for orders of integration, since the test sizes are not correct when based on a null hypothesis that is of lower order than the true DGP. These general issues are discussed by Dickey and Pantula (1987), while Ilmakunnas (1990) illustrates them in the context of seasonal unit root testing.

Typically, the highest order of integration considered for seasonal economic time series is $\Delta_1 y_t \sim SI(1)$ as implied by (3.83). Under the null hypothesis, seasonal differencing is required in addition to first differencing. Since seasonal differencing implies the presence of a zero frequency unit root (in addition to the full set of seasonal unit roots), then this null hypothesis implies the presence of two zero frequency unit roots for the levels series y_t. Following Engle, Granger, and Hallman (1988), we denote $\Delta_1 y_t \sim SI(1)$ as $y_t \sim I(2, 1)$, which is understood to indicate two zero frequency unit roots and one set of the $S - 1$ seasonal unit roots implied by the seasonal summation operator $(1 + L + \cdots + L^{S-1})$. We prefer this notation to the one suggested by Osborn et al. (1988), since the latter does not separate the zero frequency and seasonal unit root implications of the seasonal difference operator Δ_S. As indicated by our preferred notation, we do not tackle the issue (permitted in the HEGY framework) where the process may contain a subset of the seasonal unit roots. Thus, we consider only the two cases of all $S - 1$ seasonal unit roots implied by $(1 + L + \cdots + L^{S-1})$ being present or the stochastic seasonality being stationary with all roots associated with the seasonal frequencies being strictly outside the unit circle.

The overall $I(2, 1)$ null hypothesis can be examined through the test regression of Hasza and Fuller (1982), where the null hypothesis is $\phi_1 = \phi_S = 1$ and $\phi_{S+1} = -1$ in

$$y_t = \phi_1 y_{t-1} + \phi_S y_{t-S} + \phi_{S+1} y_{t-S-1} + \sum_{s=1}^{S} \gamma_s \delta_{st} + z_t.$$

The difficulty, however, is that this is a general test of the overall null hypothesis rather than a test against a specific alternative. Rejection of the null hypothesis does not, therefore, indicate a particular direction to look under the alternative hypothesis. For this reason, testing down is usually conducted against specific alternatives.

Ilmakunnas (1990) discusses the routes for unit root testing in a seasonal context. We note that three routes may be considered for the initial test of the $I(2, 1)$ null hypothesis.

1. Test against the $I(1, 1)$ alternative through a Dickey–Fuller unit root test applied to $\Delta_S y_t$, or a Dickey–Fuller test applied to $\Delta_1 y_t$, in the latter case ensuring that at least $S - 1$ lags of the dependent variable are included to take account of the possible presence of the seasonal unit roots. If the null hypothesis is rejected, then the new null hypothesis becomes $y_t \sim I(1, 1)$ and the seasonal unit root tests of Section 3.3 may then be applied to test against lower orders of integration.
2. Test against $I(2, 0)$, $I(1, 1)$, and $I(1, 0)$ alternatives through the HEGY test regression (3.64) applied to $\Delta_1 y_t$ rather than to y_t. If the null hypothesis is rejected in favor of either $I(1, 1)$ or $I(1, 0)$, then the implied zero frequency unit root has to be checked against $I(0, 1)$ or $I(0, 0)$ processes, respectively.
3. Test against $I(1, 1)$ and $I(1, 0)$ alternatives through the use of the original OCSB regression of (3.78). Once again, if the $I(2, 1)$ null hypothesis is rejected in favor of either the $I(1, 1)$ or $I(1, 0)$ alternatives, the implied zero frequency unit root still has to be checked.

Whichever of these routes is chosen, the test regression(s) used will typically require augmentation to account for autocorrelation, while appropriate deterministic terms also have to be included.

3.5 Monte Carlo Studies

Our discussion so far in this chapter has considered, almost exclusively, the theoretical asymptotic properties of the various seasonal unit root test procedures.

Clearly, however, practitioners are also interested in the finite sample performance of such procedures under various assumptions for the true data-generating process. Such examinations are almost invariably performed through the use of Monte Carlo simulations with the disturbances generated by a normal (Gaussian) distribution.

3.5.1 Comparisons of Test Procedures

GLN [Ghysels, Lee, and Noh (1994)] examines the size and power properties of the usual Dickey–Fuller (DF) test, together with the DHF and HEGY seasonal unit root tests, for various quarterly DGPs containing both stationary and nonstationary stochastic components. In general, they recommend the use of the HEGY procedure, but note that it can suffer from serious size distortions caused by the near cancellation of autoregressive unit roots with moving average roots. For example, the quarterly $SI(1)$ process

$$\Delta_4 y_t = \varepsilon_t - \theta_4 \varepsilon_{t-4}, \tag{3.84}$$

with a value of θ_4 close to (but less than) 1, leads to serious size distortions for the HEGY test statistics. In practice, using a nominal 5% level of significance, the empirical size of the test may approach unity (and hence the null hypothesis is almost always rejected) even though the null hypothesis is true. This is essentially because the moving average polynomial $(1 - \theta_4 L^4)$ results in near cancellation with $\Delta_4 = 1 - L^4$. All HEGY test statistics (at the zero, semiannual, and annual, frequencies) are affected. Although augmentation of the test regression with lags of the dependent variable improves the position, serious size distortions remain. The DHF test cannot handle this case either, because of a straightforward near cancellation. This type of size distortion has also been observed in a nonseasonal context by Schwert (1989) for unit root tests with autoregressive augmentation.

The HEGY test is also affected by the presence of a positive or negative first-order moving average term in the DGP,

$$\Delta_4 y_t = \varepsilon_t - \theta_1 \varepsilon_{t-1}, \tag{3.85}$$

with θ_1 close to $+1$ or -1. This is because the HEGY test is based on the factorization $\Delta_4 = (1 - L)(1 + L)(1 + L^2)$, so that either of the first two factors can suffer near cancellation with the MA component [see also Hylleberg (1995)]. However, GLN find that the size of the DHF test is largely unaffected by a strong first-order moving average component, because the form of the test statistic does not allow the approximate cancellation to occur.

Similarly, GLN observe that the usual DF test has poor size properties in the presence of seasonal unit roots when the regression is augmented by too few lags of $\Delta_1 y_t$. In particular, at least three lags must be included in (3.74) in a quarterly context to account for the seasonal unit roots. When appropriately augmented, this statistic has similar properties to the HEGY $t(\widehat{\pi}_1)$ statistic.

Again because of their different designs, the joint HEGY test $F(\widehat{\pi}_1 \cap \widehat{\pi}_2 \cap \widehat{\pi}_3 \cap \widehat{\pi}_4)$ has better power properties than the DHF test statistic $t(\widehat{\alpha}_4)$ against a process with deterministic seasonality and a stationary first-order autoregressive process. This is because the latter test assumes that the stationary stochastic process has a seasonal form under the alternative hypothesis. Thus, the more flexible HEGY test is preferable in this respect. Although the inclusion of redundant lags or irrelevant deterministic components may result in a loss of power, GLN recommend a conservative strategy for the inclusion of such terms (and especially seasonal dummy variables) to protect against the serious size distortions that result from their erroneous exclusion. This, of course, concurs with our general recommendation of Subsection 3.4.2 that such deterministic terms be included.

Rodrigues and Osborn (1999a) conduct a similar study to that of GLN, but using monthly data. They also include the $SI(1)$ version of the OCSB test given by (3.79). In general, their conclusions are broadly similar to those of GLN. However, they shed more light on the issue of near cancellation by showing that, when the seasonal unit root tests of DHF, HEGY, or OCSB are applied to data generated by a monthly version of (3.84), then lagrange multiplier serial correlation tests fail to indicate the need for any augmentation in the test regression. In other words, not only does near cancellation occur when the seasonal MA coefficient is close to unity (leading to all seasonal unit test statistics being very badly sized), but this misspecification is not signaled by the conventional diagnostic test for autocorrelation. Therefore, the practitioner is unlikely to realize that there is a problem with the dynamic specification being used for the test regression.

In comparing the various approaches, Rodrigues and Osborn find that the monthly HEGY test has poor power properties against stationary stochastic seasonality compared with the DHF or OCSB tests. They conclude that this is another consequence of the HEGY factorization, which not only implies the estimation of substantially more parameters in a monthly context compared with these other two tests, but also causes the autoregressive roots tested to be closer to unity. For example, the factor of $(1 - \phi_{12}L^{12})$ associated with the zero frequency is $(1 - \sqrt[12]{\phi_{12}L})$, which implies a root close to unity for moderate positive values of ϕ_{12}. The DHF and OCSB approaches, on the other hand, explicitly examine the seasonal lag coefficient.

In the context of the application of the Dickey–Fuller test to seasonally integrated data, the findings of Rodrigues and Osborn reinforce the GLN results. In particular, Dickey and Fuller find that the elimination of insignificant lags by using standard significance tests applied to the coefficients $\phi_1, \ldots, \phi_{S-1}$ of (3.74) leads to $t(\widehat{\alpha}_1)$ being badly oversized. The reason is, of course, that these standard procedures are invalid in this case as a result of the presence of the seasonal unit roots. Thus, if seasonal unit roots could be present in the data and a DF test is applied, then at least $S - 1$ lags of $\Delta_1 y_t$ must be included, irrespective of apparent insignificance according to standard criteria.

3.5.2 Augmentation, Prewhitening, and Structural Breaks

For both seasonal and nonseasonal processes, Monte Carlo studies have established that specification of the order of augmentation is crucial in practice; too little augmentation yields wrongly sized test statistics, while the inclusion of redundant lags of the dependent variable sometimes results in a substantial loss of power. One pragmatic solution often adopted in practice is the use of the data-based Akaike or Schwarz criteria for lag selection. Although not explicitly concerned with seasonal unit root tests, the findings of Hall (1994) contain some warnings about the automatic use of such criteria with seasonal series.

In particular, Hall examines the usual augmented DF test regression

$$\Delta_1 y_t = \alpha y_{t-1} + \gamma + \sum_{i=1}^{p_0} \phi_i \Delta_1 y_{t-i} + \varepsilon_t,$$

where p_0 is specified from the observed data and ε_t is assumed to be i.i.d. $(0, \sigma^2)$. When $\Delta_1 y_t$ is AR(p), and hence $\alpha = 0$, then the DF critical values remain valid asymptotically provided that $p_0 \geq p$. A procedure for specifying the order of augmentation that does not yield this asymptotically will cause the test to be inappropriately sized. Perhaps surprisingly, the seasonal quarterly DGP

$$\Delta_1 y_t = \phi_4 \Delta_1 y_{t-4} + \varepsilon_t$$

results in the augmented DF test being substantially oversized with $T = 100$ observations and selection of the augmentation lag p_0 based on an information criterion. The rationale is that the "gap" implied by the absence of autoregressive lags 1, 2, and 3 from the DGP results in such criteria's tending to specify the true lag order too low with moderate sample sizes. When the data are seasonal, the obvious solution is to consider seasonal lags separately from the intermediate ones $1, 2, \ldots, S - 1$. In any case, as we have already noted, in the context of the possible presence of seasonal unit roots, these intermediate lags have to be included in an augmented DF test regression. Once again, the implication is that a conservative lag augmentation strategy has to be adopted.

A solution to the near cancellation problem, whereby one or more components of the seasonal unit root cancels with factors in the MA part of the process, is offered by Psaradakis (1997). He proposes that seasonal unit root tests should be applied to data that are "prewhitened" under the null hypothesis. Ignoring deterministic terms for simplicity, this approach begins by applying seasonal differencing with the approximating autoregressive order p_0 specified by using a data-based method applied to the autoregression

$$\Delta_S y_t = \sum_{i=1}^{p_0} \phi_i \Delta_S y_{t-i} + \varepsilon_t. \qquad (3.86)$$

The seasonal unit root test is then applied to the prewhitened data $\tilde{y}_t = (1 - \sum_{i=1}^{p_0} \hat{\phi}_i L^i) y_t$, without augmenting with lagged variables. Psaradakis finds the sizes of the prewhitened DHF and HEGY statistics to be improved dramatically compared with the usual augmented statistics.

It is, indeed, worthy of comment that the original papers of DHF, HEGY, and OCSB all propose that two-step procedures be used to obtain their test statistics, whereby the augmentation is specified in a model imposing the null hypothesis, with the hypothesis tested in a second stage. Their procedures are based on the Taylor series approximations to the underlying process, which is assumed to be an AR(p) in $\Delta_S y_t$ or $\Delta_1 \Delta_S y_t$, as appropriate. Such two-step approaches have largely fallen out of use, with practitioners preferring to specify the order of augmentation and perform the test in a single step. However, a one-step approach effectively specifies the order of augmentation under the alternative hypothesis, because the restrictions implied by the seasonal unit root process are not imposed. The results of Psaradakis (1997) suggest that this may have undesirable consequences when the null hypothesis is true. It is an open question at the time of this writing as to whether the better properties observed by Psaradakis (1997) are essentially due to prewhitening or to the specification of the augmentation under the null rather than the alternative hypothesis.

Monte Carlo evidence about the impact of allowing for seasonal mean shifts, or structural breaks, on forecast accuracy is provided by Paap, Franses, and Hoek (1997). Perhaps not surprisingly, they find that a seasonal shift model (which allows a structural break within a deterministic seasonality specification) generally performs better than one based on a model selected by the HEGY procedure when the DGP is of seasonal mean shift type. Similarly, the HEGY procedure is typically more accurate when the DGP contains one or more seasonal unit roots, but this advantage does not apply at longer horizons. The reason for this last result is, however, unclear.

3.6   Seasonal Cointegration between Variables

The final theoretical topic of this chapter examines seasonal cointegration. When two (or more) variables contain unit roots at the seasonal frequencies, then seasonal cointegration may apply between the variables at these frequencies. Subsection 3.6.1 discusses single equation cointegration tests, while the following subsection examines this approach in a VAR (vector autoregressive) context. Although the issues in the former are met again in the latter, we nevertheless begin with a single equation analysis since it provides a simpler introduction. In line with most of our discussion in Section 3.3, we initially assume simple dynamic processes and an absence of deterministic terms in order to concentrate on the essential features of the analysis. For the same reason, we consider the quarterly case with $S = 4$.

3.6.1 A Single Equation Approach

The seminal paper of Hylleberg et al. (1990), that is HEGY, and the subsequent analysis of Engle et al. (1993) develop a seasonal cointegration methodology and apply this to the consumption function, with possible seasonal cointegration between quarterly income, y_t, and consumption, c_t. The analysis could be extended to data observed at other frequencies, although to date there are few applications to monthly data. Following HEGY and Engle et al., we assume here that we wish to model c_t as the dependent variable with y_t being weakly exogenous for the parameters of interest. Indeed, our discussion largely follows these papers. However we differ with respect to the discussion of complex unit roots, where we apply some of principles of the vector cointegration analysis of Johansen and Schaumburg (1999).

With $c_t, y_t \sim SI(1)$, the transformed variables $c_t^{(1)}$ and $y_t^{(1)}$ relate to the zero frequency (or conventional) unit root, since the summation involved in this transformation removes the seasonal unit roots. Thus, with two variables, cointegration at the zero frequency implies that there will be a unique linear combination

$$w_t^{(1)} = c_t^{(1)} - k_1 y_t^{(1)} \tag{3.87}$$

such that $w_t^{(1)} \sim I(0)$. Note that here we follow the usual normalization of setting the coefficient on the (transformed) dependent variable to unity to define the cointegrating vector. Similarly, the variables $c_t^{(2)}$ and $y_t^{(2)}$ relate to the semiannual seasonal frequency and cointegration at this frequency implies that there will be a unique stationary linear combination

$$w_t^{(2)} = c_t^{(2)} - k_2 y_t^{(2)}. \tag{3.88}$$

Turning to $c_t^{(3)}$ and $y_t^{(3)}$, we note that these variables are constructed so as to remove the real roots ± 1 from c_t and y_t, respectively, leaving the complex pair of unit roots $\pm i$ in their univariate dynamics. We can consider cointegration for the complex unit roots of $c_t^{(3)}$ and $y_t^{(3)}$ through the use of complex variables. Thus, factorize $(1 + L^2)$ as $(1 + iL)(1 - iL)$. Applying the first factor to both $c_t^{(3)}$ and $y_t^{(3)}$, we can consider cointegration beween $c_t^{(3)} + ic_{t-1}^{(3)}$ and $y_t^{(3)} + iy_{t-1}^{(3)}$. Such cointegration implies the existence of a complex coefficient $k_R - ik_I$ such that the complex variable formed as

$$c_t^{(3)} + ic_{t-1}^{(3)} - (k_R - ik_I)(y_t^{(3)} + iy_{t-1}^{(3)})$$
(3.89)

is stationary. The complex conjugate pairs of these variables must follow the corresponding relationship. Therefore,

$$c_t^{(3)} - ic_{t-1}^{(3)} - (k_R + ik_I)(y_t^{(3)} - iy_{t-1}^{(3)})$$
(3.90)

is also stationary. Adding these last two equations eliminates the complex part of the equations and yields the stationary real variable $2c_t - 2k_R y_t - 2k_I y_{t-1}$. Thus, we can define

$$w_t^{(3)} = c_t^{(3)} - k_R y_t^{(3)} - k_I y_{t-1}^{(3)}$$
(3.91)

as the stationary real variable that arises from cointegration with respect to the complex unit roots $\pm i$. Note that the variable is necessarily real and will be stationary if the complex pair of variables defined by (3.89) and (3.90) is stationary.[6]

To interpret (3.91), recall from the discussion of Subsection 3.3.3 that, because the dynamics of each of the variables $c_t^{(3)}$ and $y_t^{(3)}$ have the complex pair of roots $\pm i$, then each variable can be considered to be a nonstationary seasonal process with $S = 2$ and following separate nonstationary processes for these two seasons. Then (3.91) allows the cointegration between $c^{(3)}$ and $y^{(3)}$ to exist within the same season ($k_R \neq 0$, $k_I = 0$), across the two distinct seasons ($k_R = 0$, $k_I \neq 0$), or across a linear combination of the two seasons ($k_R \neq 0$, $k_I \neq 0$).

With cointegration at each of the zero and seasonal frequencies, the error-correction mechanism (ECM) can be written as

$$\Delta_4 c_t = \lambda_1 w_{t-1}^{(1)} + \lambda_2 w_{t-1}^{(2)} + \lambda_3 w_{t-2}^{(3)} + \lambda_4 w_{t-1}^{(3)} + z_t,$$
(3.92)

where z_t is a stationary disturbance process. Note that the inclusion of both $w_{t-2}^{(3)}$ and $w_{t-1}^{(3)}$ in this equation allows the adjustment to the cointegrating relationship of (3.91) to take place with either a one- or a two-period lag (or, indeed, both). Recalling again the two-season interpretation of the transformation embodied in $c_t^{(3)}$ and $y_t^{(3)}$, we conclude that the possibility of the adjustment taking place in either of these seasons seems intuitively plausible.

---

[6] It may appear that we have ignored the imaginary component of (3.89) and (3.90). Subtracting the latter equation from the former yields $2i(c_{t-1}^{(3)} + k_I y_t^{(3)} - k_R y_{t-1}^{(3)})$. However, $k_I y_t^{(3)}$ can be replaced in this cointegrating relationship by $-k_I y_{t-2}^{(3)}$ as $y_t^{(3)} = -y_{t-2}^{(3)} + (1 + L^2)y_t^{(3)}$ with $(1 + L^2)y_t^{(3)} = \Delta_4 y_t$ stationary. Thus, analysis of the complex part also implies that $w_t^{(3)}$ defined by (3.91) is stationary.

Substitution from (3.87), (3.88), and (3.91) into (3.92) yields

$$
\begin{aligned}
\Delta_4 c_t &= \lambda_1 \left[c_{t-1}^{(1)} - k_1 y_{t-1}^{(1)} \right] + \lambda_2 \left[c_{t-1}^{(2)} - k_2 y_{t-1}^{(2)} \right] \\
&\quad + \lambda_3 \left[c_{t-2}^{(3)} - k_R y_{t-2}^{(3)} - k_I y_{t-3}^{(3)} \right] \\
&\quad + \lambda_4 \left[c_{t-1}^{(3)} - k_R y_{t-1}^{(3)} - k_I y_{t-2}^{(3)} \right] + z_t \\
&= \lambda_1 \left[c_{t-1}^{(1)} - k_1 y_{t-1}^{(1)} \right] + \lambda_2 \left[c_{t-1}^{(2)} - k_2 y_{t-1}^{(2)} \right] \\
&\quad + \left[\lambda_3 c_{t-2}^{(3)} - (\lambda_3 k_R + \lambda_4 k_I) y_{t-2}^{(3)} \right] \\
&\quad + \left[\lambda_4 c_{t-1}^{(3)} - (\lambda_4 k_R - \lambda_3 k_I) y_{t-1}^{(3)} \right] + z_t^*,
\end{aligned}
\tag{3.93}
$$

where $z_t^* = z_t - \lambda_3 k_I (1 + L^2) y_{t-1}^{(3)}$ is stationary. Note that the rearrangement to give the second form here includes the use of the definitional relationship $y_{t-3}^{(3)} = (1 + L^2) y_{t-1}^{(3)} - y_{t-1}^{(3)}$. The component $(1 + L^2) y_{t-1}^{(3)} = \Delta_4 y_{t-1}$ is stationary by construction. The ECM as written in (3.93) is convenient because it isolates the transformed variables $c^{(3)}$ and $y^{(3)}$ appearing at lag 2 from those at lag 1. However, its behavioral interpretation is less obvious than that of the ECM as written in (3.92). Indeed, as argued in Osborn (1993), the transformed variables in (3.93) could be substituted in terms of $c_{t-1}, \ldots, c_{t-4}, y_{t-1}, \ldots, y_{t-4}$ to yield adjustment relationships written as functions of observed variables at each of the lags $1, \ldots, 4$. The implied adjustments then differ with the lag, but it is again difficult to see the underlying behavioral interpretation.

The two-step cointegration method of Engle and Granger (1987) can be applied to obtain an estimated error-correction model for the consumption function. Having tested for seasonal unit roots in each of c_t and y_t by using the HEGY approach, the first stage is to examine cointegration at each of the zero, semiannual, and annual frequencies by estimating the coefficients of (3.87), (3.88), and (3.91) by OLS regressions. Again as a result of the asymptotic orthogonality of the HEGY transformed variables, separate estimation of these relations is valid.

The residuals from the first regression can be tested for stationarity by using the Dickey–Fuller test

$$
\Delta_1 \widehat{w}_t^{(1)} = \pi_1 \widehat{w}_{t-1}^{(1)} + \varepsilon_t,
\tag{3.94}
$$

where residuals from the estimated cointegrating relationship are indicated by $\widehat{w}_t$. Modified critical values, such as those of MacKinnon (1991), have to be used to take account of the prior estimation of the cointegrating relationship. Testing cointegration at the semiannual frequency is a test for whether the unit root of -1 in both c_t and y_t no longer occurs in the linear combination $w_t^{(2)}$. By analogy with (3.66), and by using the estimated version of (3.88), this involves a test of $\pi_2 = 0$ against $\pi_2 < 0$ in

$$\left(\widehat{w}_t^{(2)} + \widehat{w}_{t-1}^{(2)}\right) = -\pi_2 \widehat{w}_{t-1}^{(2)} + \varepsilon_t. \tag{3.95}$$

As discussed in Subsection 3.3.4 in the context of observed variables, $t(\widehat{\pi}_2)$ again follows a Dickey–Fuller type distribution.

In a similar manner, the test for cointegration at the annual frequency is a test for whether the linear combination $w_t^{(3)}$ does not contain the complex pair of unit roots at this frequency. Again by analogy with our discussion of the HEGY test, and in particular (3.67), this implies a test of $\pi_3 = \pi_4 = 0$ in

$$\widehat{w}_t^{(3)} + \widehat{w}_{t-2}^{(3)} = -\pi_3 \widehat{w}_{t-2}^{(3)} - \pi_4 \widehat{w}_{t-1}^{(3)} + \varepsilon_t. \tag{3.96}$$

Once again, the critical values used for testing the coefficients in (3.96) have to take account that the variables are formed as residuals from an estimated cointegrating relationship. Engle et al. (1993) obtain appropriate critical values by Monte Carlo simulation. However the univariate analysis of Burridge and Taylor (2000) for the pair of complex unit roots discussed in Subsection 3.4.1 above, indicates that the coefficients in (3.96) should be tested jointly.

Assuming that cointegration is found for each frequency, the second stage would be to estimate the error-correction model in the form of (3.92), using lagged residuals from the first step in place of the quantities $w_{t-1}^{(1)}$, $w_{t-1}^{(2)}$, $w_{t-2}^{(3)}$, and $w_{t-1}^{(3)}$. For any frequency in which there is no cointegration, corresponding term(s) will be omitted.

3.6.2   The Vector Approach

Since cointegration is essentially a multivariate concept, it is natural to examine cointegration issues in an explicitly multivariate framework. The VAR approach to nonseasonal cointegration of Johansen (1988, 1991) is extended to the seasonal case by Lee (1992). That analysis is further extended and also corrected by Johansen and Schaumburg (1999). At the time of writing, the latter paper must be regarded as the primary reference on the econometrics of seasonal cointegration.

For simplicity of exposition, we initially assume that the VAR process for the $N \times 1$ vector of seasonal time series Y_t observed S times per year is of order S, so that

$$Y_t = \Phi_1 Y_{t-1} + \cdots + \Phi_S Y_{t-S} + U_t, \tag{3.97}$$

where U_t is a vector i.i.d. process with $E(U_t U_t') = \Sigma$. It is important to appreciate that the vector Y_t in the present context is a vector of N series observed at time period t, with U_t the corresponding $N \times 1$ disturbance vector. These should not be confused with the vectors Y_τ and U_τ used earlier, where these relate to the S values on a single series for the seasons of year τ. Similarly, when we refer to an element y_{it} in the current section, this refers to the ith series observed at time t. In common with our univariate discussion of seasonal unit roots, we initially make the unrealistic assumptions that the starting values $Y_{-S+1} = \cdots = Y_0 = 0$ and that the process (3.97) contains no deterministic component. These assumptions ensure that $E(Y_t) = 0$ for all t. We also assume that each of the elements of the vector Y_t is seasonally integrated, so that all series are individually $SI(1)$.

Seasonal cointegration considers the ECM, for (3.97). This ECM is derived from

$$\Delta_S Y_t = \Phi_1 Y_{t-1} + \cdots + \Phi_{S-1} Y_{S-1} + (\Phi_S - I_S)Y_{t-S} + U_t. \quad (3.98)$$

However, rather than the variables $Y_{t-1}, \ldots, Y_{t-S}$, the analysis considers linear combinations of these vectors that separate the unit roots at the zero and each of the seasonal frequencies. In particular, generalizing (3.16), we can define

$$Y_t^{(1)} = Y_t + Y_{t-1} + Y_{t-2} + Y_{t-3}, \quad (3.99)$$

$$Y_t^{(2)} = Y_t - Y_{t-1} + Y_{t-2} - Y_{t-3}, \quad (3.100)$$

$$Y_t^{(3)} = Y_t - Y_{t-2}. \quad (3.101)$$

The ECM considered by Johansen and Schaumburg can then be written as

$$\Delta_4 Y_t = \Pi_1 Y_{t-1}^{(1)} + \Pi_2 Y_{t-1}^{(2)} + \Pi_3 Y_{t-2}^{(3)} + \Pi_4 Y_{t-1}^{(3)} + U_t. \quad (3.102)$$

Except for (some) signs, this representation is directly analogous to the HEGY equation (3.64).

Within this approach, and because of the asymptotic orthogonality of the regressors in the ECM, cointegration restrictions can be investigated separately for each of the zero frequency and seasonal unit roots. The first two error-correction terms in (3.102) can be interpreted fairly readily from our discussion above. Thus, each of the elements in the vector $Y_t^{(1)}$ is a conventional $I(1)$ variable because the seasonal unit roots have been eliminated in (3.99) through the transformation $(1 + L + L^2 + L^3)$. The matrix coefficient $\Pi_1 = \Lambda_1 K_1'$ can be obtained through a vector cointegration analysis applied to $Y_t^{(1)}$. That is, if the matrix Π_1 has rank r_1 with $0 < r_1 < N$, then there are r_1 cointegrating relationships between the variables at the zero frequency. These cointegrating relationships are given by the columns of the $N \times r_1$ matrix K_1, while the elements of the $N \times r_1$ matrix Λ_1 yield the adjustment coefficients. Similarly, if the matrix $\Pi_2 = \Lambda_2 K_2'$ has rank r_2 with $0 < r_2 < N$, then there are r_2 cointegrating relationships between the variables at the semiannual frequency (recall that the transformation embodied in $Y_t^{(2)}$ removes the zero and annual frequency unit roots). The r_2 semiannual cointegrating relationships are given by the columns of the $N \times r_2$ matrix K_2, with adjustment coefficients given by the $N \times r_2$ matrix Λ_2.

To concentrate on the complex unit roots, assume for the moment that $\Pi_1 = \Pi_2 = 0$ in (3.102). Then this ECM would become

$$(1 + L^2)Y_t^{(3)} = \Pi_3 Y_{t-2}^{(3)} + \Pi_4 Y_{t-1}^{(3)} + U_t$$

$$= (\Pi_R + i\Pi_I)X_{t-1}^{(1)} + (\Pi_R - i\Pi_I)X_{t-1}^{(2)} + U_t.$$

The first line here is analogous to the univariate equation for a complex pair of unit roots, namely (3.60). In line with the discussion of Subsection 3.3.3, the second line is obtained by explicitly considering the complex roots by defining the vector pair of complex variables $X_t^{(1)} = Y_t^{(3)} + iY_{t-1}^*$ and $X_t^{(2)} = Y_t^{(3)} - iY_{t-1}^{(3)}$, which have complex (matrix) coefficients of $\Pi_R \pm i\Pi_I$ in the ECM. If these complex coefficient matrices each have reduced rank r_c, where $0 < r_c < N$ (as a complex pair they must have the same rank), then this ECM becomes

$$(1 + L^2)Y_t^{(3)} = (\Lambda_R + i\Lambda_I)(K_R - iK_I)'X_{t-1}^{(1)}$$

$$+ (\Lambda_R - i\Lambda_I)(K_R + iK_I)'X_{t-1}^{(2)} + U_t, \qquad (3.103)$$

where the complex matrices $\Lambda_R + i\Lambda_I$ and $K_R + iK_I$ are each $N \times r_c$ and of rank r_c. Once again, the cointegrating relationships are given by the columns of $K_R \pm iK_I$ while $\Lambda_R \pm i\Lambda_I$ yield the adjustment coefficients. This ECM can be converted into one involving only real values by substituting the definitions for $X_{t-1}^{(1)}$ and $X_{t-1}^{(2)}$, which then yields

$$(1 + L^2)Y_t^{(3)} = 2(\Lambda_R K_R' + \Lambda_I K_I')Y_{t-1}^{(3)} + 2(\Lambda_R K_I' - \Lambda_I K_R')Y_{t-2}^{(3)} + U_t.$$

$$(3.104)$$

This equation is a generalization of the second line of the univariate equation (3.63) concerning the complex unit roots $\pm i$.

Returning to the ECM of (3.102), we see that allowing all coefficients to be nonzero and substituting in terms of the cointegrating relationships yields

$$\Delta_4 Y_t = \Lambda_1 K_1' Y_{t-1}^{(1)} + \Lambda_2 K_2' Y_{t-1}^{(2)} + (\Lambda_R + i\Lambda_I)(K_R - iK_I)'X_{t-1}^{(1)}$$

$$+ (\Lambda_R - i\Lambda_I)(K_R + iK_I)'X_{t-1}^{(2)} + U_t,$$

$$= \Lambda_1 K_1' Y_{t-1}^{(1)} + \Lambda_2 K_2' Y_{t-1}^{(2)} + 2(\Lambda_R K_I' - \Lambda_I K_R')Y_{t-2}^{(3)}$$

$$+ 2(\Lambda_R K_R' + \Lambda_I K_I')Y_{t-1}^{(3)} + U_t. \qquad (3.105)$$

From the perspective of a behavioral equation, (3.105) does not appear to have a ready interpretation in respect of the complex unit roots, since each matrix coefficient $2(\Lambda_R K_I' - \Lambda_I K_R')$ and $2(\Lambda_R K_R' + \Lambda_I K_I')$ involves both the cointegrating coefficient matrices (K_R, K_I) and both the adjustment matrices

(Λ_R, Λ_I). In their corresponding discussion, Johansen and Schaumburg consider some possible simplifications, which have testable implications. However, our discussion of the single-equation case in Subsection 3.6.1 appears to throw some light on this issue. In particular, the form of the vector ECM in (3.105) is analogous to the form we obtained in the single-equation case as (3.93). To aid interpretation, however, we preferred the use of the form (3.92).

Applied in this vector context, therefore, our proposal to aid interpretation is to consider the cointegrating linear combinations $(K_R - iK_I)'X_t^{(1)} = (K_R - iK_I)'(Y_t^{(3)} + iY_{t-1}^{(3)})$ and $(K_R + iK_I)'X_t^{(2)} = (K_R + iK_I)'(Y_t^{(3)} - iY_{t-1}^{(3)})$ through the combined cointegrating relationship $K_R'Y_t^{(3)} + K_I'Y_{t-1}^{(3)}$ involving only real variables. The need to consider cointegration over two time periods again follows from the implication of the complex pair of roots that the process $Y_t^{(3)}$ fundamentally consists of two separate (vector) processes relating to t and $t - 1$. Then, rather than (3.105), we can represent the ECM as

$$
\begin{aligned}
\Delta_4 Y_t &= \Lambda_1 K_1' Y_{t-1}^{(1)} + \Lambda_2 K_2' Y_{t-1}^{(2)} + \Lambda_3 \big(K_R' Y_{t-2}^{(3)} + K_I' Y_{t-3}^{(3)} \big) \\
&\quad + \Lambda_4 \big(K_R' Y_{t-1}^{(3)} + K_I' Y_{t-2}^{(3)} \big) + U_t, \\
&= \Lambda_1 K_1' Y_{t-1}^{(1)} + \Lambda_2 K_2' Y_{t-1}^{(2)} + \Lambda_3 \big(K_R' Y_{t-2}^{(3)} - K_I' Y_{t-1}^{(3)} \big) \\
&\quad + \Lambda_4 \big(K_R' Y_{t-1}^{(3)} + K_I' Y_{t-2}^{(3)} \big) + U_t^*, \qquad (3.106)
\end{aligned}
$$

where $U_t^* = U_t + \Lambda_3 K_I'(Y_{t-1}^{(3)} + Y_{t-3}^{(3)})$ is stationary.

The behavioral interpretation of the first form in (3.106) is that adjustment to this cointegrating relationship can take place with a lag of two periods or one period, with the respective adjustment coefficients being Λ_3 and Λ_4. The second form replaces Y_{t-3}^* by $Y_{t-1}^{(3)}$ and, by comparison with (3.105), it follows that $\Lambda_3 = -2\Lambda_I$ and $\Lambda_4 = 2\Lambda_R$. Fundamentally, therefore, both Λ_3 and Λ_4 are adjustment matrices for the cointegrating relationships at the annual seasonal frequency, with Λ_3 capturing adjustment at lag 2 and Λ_4 adjustment at lag 1.

Despite this progress, however, seasonal cointegration is difficult to motivate from a behavioral perspective, and the comments made in the previous subsection apply in this vector case also.

We will not here go into the technical aspects of estimation of the ECM or the asymptotic properties of the estimators; the interested reader is referred to the original paper by Johansen and Schaumburg (1999) for this. However, it may be noted that the essential feature is that, when a HEGY-type approach is taken, issues relating to cointegration at the zero and each of the seasonal frequencies can effectively be treated separately. The important distinctive issues that arise are then concerned with cointegration at those seasonal frequencies that correspond to complex pairs of seasonal unit roots. For the real unit roots of $+1$ and -1, the procedures follow the reduced rank regression approach underlying nonseasonal cointegration; see Johansen and Schaumburg (1999) and Lee (1992). To deal with the complex pair of unit roots, Johansen and Schaumburg define expanded matrices that contain both the real and complex parts of the coefficient matrices or variables, as required. This representation essentially allows a generalized form of cointegration analysis to be applied at this frequency.

3.6.3 Seasonal Intercepts and Seasonal Cointegration

As discussed in Subsection 3.2.2, the presence of seasonal intercept terms in a seasonally integrated process implies that the mean of the variable follows a linear trend for each of the seasons of the year. If these intercepts are distinct over the S seasons, then so also are the trends followed by the series $y_{s\tau}$ over $s = 1, \ldots, S$. Extending the analysis of Johansen (1994) for nonseasonal cointegration, Franses and Kunst (1999) examine the role of seasonal intercepts within seasonal cointegration. Their analysis is based on quarterly data and the ECM of (3.102) with the addition of seasonal intercepts; this analysis is also used by Johansen and Schaumburg (1999). Thus, consider the ECM

$$\Delta_4 Y_t = \sum_{s=1}^{4} \Gamma_s \delta_{st} + \Pi_1 Y_{t-1}^{(1)} + \Pi_2 Y_{t-1}^{(2)} + \Pi_3 Y_{t-2}^{(3)} + \Pi_4 Y_{t-1}^{(3)} + U_t,$$

(3.107)

where Γ_s is an $N \times 1$ vector of constants relating to season s and δ_{st} is, as previously, a seasonal dummy variable.

If $\Pi_1 = \Pi_2 = \Pi_3 = \Pi_4 = 0$ in (3.107), then it is clear from our discussion of Subsection 3.2.2 that $E(Y_t) = E(Y_{s\tau})$ will follow the distinct linear trends $E(Y_{s0}) + \Gamma_s \tau$ over $s = 1, \ldots, S$ when Γ_s varies over the seasons. As Franses and Kunst establish, this continues to be the case in the presence of seasonal cointegration, unless the seasonally varying intercepts are restricted to be part of the cointegrating relationship(s). They propose a simple modification of the seasonal cointegration ECM to implement these restrictions.

In (3.107), the seasonal dummy variable component $\sum_{s=1}^{*} \Gamma_s \delta_{st}$ can be equivalently reparameterized through the trigonometric representation of deterministic seasonality as $\sum_{s=1}^{4} B^{(s)} \delta_t^{(s)}$, where the new seasonal dummy variables are defined through the transformation (3.16). Thus, $\delta^{(1)} = \sum_{s=1}^{4} \delta_{st}$, $\delta_t^{(2)} = \delta_{4t} - \delta_{3t} + \delta_{2t} - \delta_{1t}$, $\delta_t^{(3)} = \delta_{4t} - \delta_{2t}$, and $\delta_t^{(4)} = -\delta_{3t} + \delta_{1t}$. Referring back to the discussion of the trigonometric representation in Subsection 2.2.2 of Chapter 2, we find that $\delta^{(1)}$ is a conventional constant that always takes the value unity (and hence the time subscript is redundant), $\delta_t^{(2)}$ follows a semi-annual cycle with values that alternate between -1 and $+1$ each period, and $\delta_t^{(3)}$ and $\delta_t^{(4)}$ both follow annual cycles as $(0, -1, 0, +1)$ and $(+1, 0, -1, 0)$, respectively. To capture annual cycles covering all four periods, both $\delta_t^{(3)}$ and $\delta_t^{(4)}$ are required, or alternatively $\delta_t^{(3)}$ and $\delta_{t-1}^{(3)}$ may be used as $\delta_t^{(4)} = -\delta_{t-1}^{(3)}$. As explained in Subsection 3.2.4, the analogy with the definitions of the HEGY variables is no coincidence.

With appropriate scaling, the coefficient vectors $B^{(s)}$ are defined in a comparable way to the dummy variables $\delta_t^{(s)}$. In particular, $B^{(1)} = (1/4) \sum_{s=1}^{4} \Gamma_s$, $B^{(2)} = (1/4)(\Gamma_4 - \Gamma_3 + \Gamma_2 - \Gamma_1)$, $B^{(3)} = (1/2)(\Gamma_4 - \Gamma_2)$, and $B^{(4)} = (1/2)(\Gamma_1 - \Gamma_3)$. It is easily verified that $\sum_{s=1}^{4} \Gamma_s \delta_{st} = \sum_{s=1}^{4} B^{(s)} \delta_t^{(s)}$. Indeed, this is a consequence of the equivalence between the dummy variable and trigonometric representations of deterministic seasonality discussed at length in Chapter 2, with $\sum_{s=1}^{4} \Gamma_s \delta_{st}$ being the dummy variable representation and $\sum_{s=1}^{4} B^{(s)} \delta_t^{(s)}$ the trigonometric one.

Now, consider cointegration at the zero frequency. If there are r_1 cointegrating relationships between the variables of $Y_t^{(1)}$, each of these r_1 relationships may include a constant, giving rise to the zero-mean stationary cointegrating relationships $K_1' Y_t^{(1)} + G_1 \delta^{(1)}$. In this case, G_1 is an $r_1 \times 1$ vector of constants. Similarly considering the cointegrating relationships for $Y_t^{(2)}$, the zero-mean cointegrating relationships become $K_2' Y_t^{(2)} + G_2 \delta_t^{(2)}$, with G_2 being $r_2 \times 1$. As argued in Subsection 3.6.2, the appropriate cointegrating relationship for the complex pair of unit roots is $K_R' Y_t^{(3)} + K_I' Y_{t-1}^{(3)}$. Extending this to allow for a constant in each of the r_c cointegrating relationships leads to $K_R' Y_t^{(3)} + K_I' Y_{t-1}^{(3)} + G_3 \delta_t^{(3)} + G_4 \delta_{t-1}^{(3)}$, where G_3 and G_4 are $r_c \times 1$ vectors (recall that both $\delta_t^{(3)}$ and $\delta_{t-1}^{(3)}$ are required to cover the four seasons). Thus, with seasonal cointegration and with the seasonal dummy variables restricted to arise in (3.107) only through these cointegrating relationships, then by generalization of the first form of (3.106), the ECM becomes

$$\Delta_4 Y_t = \Lambda_1 \left(K_1' Y_{t-1}^{(1)} + G_1 \right) + \Lambda_2 \left(K_2' Y_{t-1}^{(2)} + G_2 \delta_{t-1}^{(2)} \right)$$
$$+ \Lambda_3 \left(K_R' Y_{t-2}^{(3)} + K_I' Y_{t-3}^{(3)} + G_3 \delta_{t-2}^{(3)} + G_4 \delta_{t-3}^{(3)} \right)$$
$$+ \Lambda_4 \left(K_R' Y_{t-1}^{(3)} + K_I' Y_{t-2}^{(3)} + G_3 \delta_{t-1}^{(3)} + G_4 \delta_{t-2}^{(3)} \right) + U_t.$$

$$(3.108)$$

Alternatively, it may be more realistic to allow the VAR to include an unrestricted overall constant vector Γ, which is not seasonally varying. As an unrestricted vector, this will have dimension $N \times 1$. This allows the possibility that the process Y_t may have a common drift component but not divergent seasonal drifts; see Franses and Kunst (1999) or Johansen and Schaumburg (1999). In this case, $\Lambda_1 G_1$ is replaced in (3.108) by Γ, with the ECM otherwise unchanged.

As is usual with tests for the cointegration rank, the inclusion of deterministic components and whether they are restricted or not influences the relevant critical values. Johansen and Schaumburg (1999) provide critical values relevant to the complex unit root. See also Franses and Kunst (1999), who further investigate various additional practical issues in this context, including the augmentation of the test regression with lagged $\Delta_4 Y_t$. It should be noted, however, that these authors consider only a restricted form of (3.107) with $\Pi_4 = 0$, which then implies the restrictions $K_I = \Lambda_4 = 0$, or alternatively $K_R = \Lambda_3 = 0$, in (3.108).

3.7   Some Remarks on Empirical Results

Although it is not our intention to survey the substantial empirical literature relating to seasonal unit roots, it is appropriate to make some remarks on the broad pattern of empirical results obtained to date. There have been many applications of the univariate tests of Section 3.3 , particularly of the HEGY test approach. There is, however, no consensus among the authors of the various studies about how common seasonal unit roots are in observed macroeconomic series.

Perhaps most prominently, Miron and his colleagues find relatively little evidence in favor of seasonal unit roots in quarterly or monthly series across a range of countries; this evidence is summarized in Miron (1996). In contrast, based on an analysis of some data used in the Miron studies, Hylleberg et al. (1993) find that "a varying and changing seasonal pattern is a common phenomenon." Hylleberg (1994) attributes these different results to the use of a relatively low order of augmentation in the Miron studies, with near cancellation effects causing the seasonal unit root null hypothesis to be rejected too frequently (see the discussion of Section 3.5). Nevertheless, it might also be noted that, in common with nearly all empirical studies of seasonal unit roots, the Hylleberg et al. study considers the seasonal integration null hypothesis through a sequence of tests by conducting separate tests for the zero frequency and each seasonal frequency. In particular, no control is made of the overall level of significance. Further, relatively few of the series they examine are found to contain all of the S seasonal unit roots implied by seasonal integration.

The quarterly real GDP data for fourteen countries used by Hylleberg et al. (1993) are analyzed again by Franses and Vogelsang (1998) from the perspective of possible structural breaks in the deterministic components (as in Subsection 3.4.3). Their conclusions depend crucially on the treatment of the structural break examined (in particular, the use of an additive outlier versus an innovative outlier model). It is also the case that the apparent evidence for seasonal unit roots sometimes disappears when a structural break is allowed.

Another type of empirical evidence relates to forecast accuracy. There are, however, few studies that examine whether (or not) empirical testing for seasonal unit roots results in models with more accurate postsample forecasts. Perhaps the most extensive study (in terms of the number of series examined) that pursues this line is that of Osborn, Heravi, and Birchenhall (1999), who consider two-digit industrial production series for three European countries. They find that, despite the series typically providing evidence against seasonal integration, models based on seasonal differences produce forecasts that are at least as accurate overall (in terms of root-mean-square forecast error) as those based on deterministic seasonality. Although forecasts from the latter might be improved by allowing for structural breaks, as examined in a relatively small-scale forecast comparison exercise by Paap et al. (1997), it is also the case that (as Paap et al., note) structural change models cannot forecast these breaks when they occur during the forecast period.

Empirical studies of seasonal cointegration have also been undertaken. For example, based on their two-step approach, Engle et al. (1993) find little evidence of seasonal cointegration for consumption and income in Japan. Kunst (1993) examines seasonal cointegration between major macroeconomic variables for a number of European countries and finds evidence of cointegration at the seasonal frequencies. A similar conclusion is reached for U.K. consumption and income in Hylleberg et al. (1990), although this latter paper considers possible cointegration only through the specific linear combination $c_t - y_t$. Other studies are summarized by Franses and McAleer (1998). As yet, however, few empirical "stylized facts" have emerged with respect to the existence of seasonal cointegration. Given the unresolved debate about the prevalence of seasonal unit roots in a univariate context, this seems hardly surprising.

It was noted in Subsection 3.4.2 that deterministic terms play an important practical role in unit root tests in delivering invariance of the test statistics to nuisance parameters; Subsection 3.6.3 considered the corresponding issue in the context of cointegration. Perhaps surprisingly, the empirical literature has paid little attention to whether such terms should be included when the issue is forecast accuracy, rather than testing the unit roots. One exception to this lack of attention is Kunst and Franses (1998), who consider the role of a constant and seasonal dummy variables when forecasting using a (seasonal) cointegrated system. They find that, in general, restricting the seasonal dummies to enter only through the cointegrating relationships, as in (3.108), yields more accurate forecasts than the inclusion of unrestricted dummies.

To date, empirical studies of seasonal unit roots (in either a univariate or multivariate setting) have typically lacked an economic rationale for the presence of such unit roots. As pointed out by Osborn (1993) and noted in Subsection 3.2.1 above, seasonal integration implies a lack of cointegration between the S series $y_{s\tau}$ ($s = 1, \ldots, S$). A question rarely asked is: What economic forces might lead to such a lack of cointegration between the series for the S seasons of the year? Or, from a slightly different perspective, it is not clear what sort of optimizing behavior by consumers would lead them to adjust to the seasonal cointegrating relationships for consumption and income that arise from a consideration of the HEGY variables. Nevertheless, the other extreme of unchanging deterministic seasonality is implausible in the long run. With economic growth and evolving institutional arrangements, it seems preferable to allow that underlying preferences and technology may evolve over time, and such evolution may have a seasonal aspect. Thus, for instance, holidays traditionally taken by workers in, say, the month of August may now be taken in June, July, or September as a result of a mixture of consumers' preferring to spread their holidays and technological changes that no longer imply that it makes economic sense to close a factory for a period during which all employees take their annual vacation. When underlying economic forces for changing seasonality are considered, the implied evolution may involve adjacent quarters or nearby months, rather than one in which "summer can become winter." Such arguments might lead one to consider a methodology that anchors the analysis to specific seasons through the explicit use of the S series $y_{s\tau}$ $(s = 1, \ldots, S)$, rather than using an approach (such as HEGY) that is based on transformations convenient for their statistical properties. The former type of analysis is pursued in Chapter 6, where periodic processes are examined.

Nevertheless, whatever conclusion is reached about the occurrence of seasonal integration for observed economic time series in practice, methodologies are now available for testing the null hypothesis of deterministic seasonality (Chapter 2) and of seasonal integration (this chapter). Thus, there is no reason for automatically assuming that nonstationarity in a seasonal time series should be removed by taking annual differences or that seasonally adjusted data are required for economic modeling. Despite this, practitioners (in general) continue to base their models on seasonally adjusted data without questioning what seasonal unit root (or other) assumptions this adjustment implicitly makes. This not only raises the methodological issue of how seasonal adjustment is undertaken by official agencies (discussed in Chapter 4), but also what pitfalls may be in store for the practitioner as a result of using adjusted data (considered in Chapter 5).

4　Seasonal Adjustment Programs

4.1　Introduction

Statistical agencies around the globe engage in a process called seasonal adjustment. Roughly speaking, this amounts to filtering raw data such that seasonal fluctuations disappear from the series. Neither the principle nor its practical implementation is simple. Many resources are constantly devoted to the design of seasonal adjustment procedures. It is a subject of perpetual debate, sometimes even involving heads of state.[1] Most statistical agencies use fairly standardized procedures. The U.S. Census Bureau X-11 method and its recent upgrade, the X-12-ARIMA program, are the most widely used. A competing procedure, called TRAMO/SEATS, was developed by Agustin Maravall of the Bank of Spain. There are also other programs besides X-12 and TRAMO/SEATS, such as the procedure in Andrew Harvey's STAMP program. We will cover those in less detail, as the bulk of seasonal adjustments use either one of the two leading programs we discuss in this chapter.

The chapter is divided into five sections. The first discusses decompositions, followed by three sections devoted to the X-11, X-12-ARIMA, and TRAMO/SEATS procedures, respectively. The final section covers the subject of seasonal adjustment and other transformations.

4.2　Decompositions

Observed economic time series can be decomposed in several mutually orthogonal unobserved components. As we noted in Chapter 1, there is a long tradition dating back to the 19th Century of unobserved component models. Any seasonal adjustment procedure rests on a specific decomposition of a series into a trend

---

[1] In his ET Interview, James Durbin (*Econometric Theory* 4, 1988, pp. 140–141) refers to a situation about seasonal adjustment of key macroeconomic data involving the British Prime Minister.

cycle, and seasonal and irregular components. Typically a series y_t is decomposed into the *product* of a trend cycle y_t^{tc}, seasonal y_t^s, and irregular y_t^i, yielding

$$y_t \equiv y_t^{tc} \times y_t^s \times y_t^i. \tag{4.1}$$

Such a multiplicative decomposition applies to the majority of macroeconomic time series as it is appropriate for series of positive values in which the size of the seasonal oscillations increases with the level of the series. Seasonal adjustment in such cases amounts to dividing y_t by an estimate of y_t^s. This estimate of y_t^s is usually called the *seasonal factor*. In many circumstances the seasonal component is assumed to be stochastic and augmented by deterministic effects due to the length of months, the number of trading days, and holidays (such as Easter, Labor Day, Thanksgiving, Christmas, etc.). Summarizing these deterministic effects as y_t^{td} (trading day, including length of months) and y_t^h yields a more general multiplicative decomposition:

$$y_t = y_t^{tc} \times y_t^s \times y_t^{td} \times y_t^h \times y_y^i. \tag{4.2}$$

To simplify the discussion it will often be assumed that y_t^{td} and y_t^h are either absent or have been treated separately. The preadjustment of series for y_t^{td} and y_t^h is in fact often done in practice, as we will discuss later.

The multiplicative decomposition (4.1), or its more general form (4.2), is not always appropriate. There are three other decompositions commonly encountered. These are: (1) additive, (2) log additive, and (3) pseudo-additive decompositions. The last of these is defined as

$$y_t = y_t^{tc}\left(y_t^s + y_t^i - 1\right) = y_t^{tc} \times \left(y_t^s - 1\right) + y_t^{tc} \times y_t^i \tag{4.3}$$

and has been used at the U.K. Office for National Statistics, where it was developed to adjust series that have small, possibly zero, values in the same month(s) each year. Such series include, for instance, agricultural sector produce only available during part of the year. The multiplicative decomposition does not work well in such cases, as some seasonal factors are small, yielding large and erratic behavior of y_t/y_t^s.

The additive and log-additive decompositions are the same except that the former applies to y_t whereas the latter applies to log y_t. The additive decomposition is defined as

$$y_t = y_t^{tc} + y_t^s + y_t^i. \tag{4.4}$$

This decomposition has been the focus of most of the academic research. Usually it is assumed that y_t represents a logarithmic transformation of the raw data and, hence, it is as if one considers the multiplicative decomposition (4.1). However, a program such as X-11 has multiplicative and log-additive versions that are not identical. Consequently, applying the multiplicative version of the X-11 program or applying the additive X-11 to the logs and taking the exponential of the output does not result in the same seasonal adjustment. The log-additive decomposition requires a bias correction for its trend estimates, as a result of geometric means being less than arithmetic means [discussed by Thomson and Ozaki (1992)] as well as different calibration for extreme value identification (more on this later).

Of course, we have not yet defined what time series structure and identification is imposed on each of the latent components. Each seasonal adjustment program assumes either implicitly or explicitly structures for the y_t^{tc}, y_t^h, y_t^{td}, y_t^s, and y_t^i components. We begin with the X-11 program, which is the least explicit about the time series structure of these components.

4.3 The X-11 Program

The Census X-11 program is considered first since it is the most widely applied adjustment procedure.[2] It was the product of several decades of research. Its development started in the early 1930s by researchers at the N.B.E.R. [see, e.g., Macaulay (1931)], and it emerged as a fully operational procedure in the mid-1960s as a result of the work by Julius Shiskin and his collaborators at the U.S. Bureau of the Census [see Shiskin, Young, and Musgrave (1967)]. The program consists of a set of moving average filters that are applied to the data sequentially. The X-11 program can be approximated by a linear filter when the default option is considered. Several authors, including Wallis (1974), Laroque (1977), Dagum (1983), Burridge and Wallis (1984), Ghysels (1984), and Bell (1992), have studied the quarterly and/or monthly linear X-11 filters. The structure of the linear approximation for the quarterly filter is presented in Laroque (1977). The actual filter weights of the linear approximation to the monthly filter appear in Ghysels and Perron (1993).[3] Since the monthly filter is used more often than the quarterly one, we pay more attention to the former. In the first subsection we review the linear filter representation; the second subsection covers potential sources of nonlinearity in the X-11 program.

4.3.1 The Linear X-11 Filter

We will consider the decompositions discussed in the previous section and denote them by M for the multiplicative (4.1), PA for the pseudo-additive (4.3),

---

[2] Only symmetric filters will be discussed in this section. Hence we do not cover the X-11 ARIMA variant [Dagum (1982)] of the Census X-11 seasonal adjustment procedure. It will be covered in the next section, where X-12-ARIMA is considered.

[3] Several authors, including Dagum (1983) and Burridge and Wallis (1984), graph the transfer function of the linear monthly filter. Burridge and Wallis tabulated some of the filter weights. In the Appendix to Ghysels and Perron (1993), all filter weights for the monthly and quarterly linear filters are tabulated.

and A for the (log-) additive. The different steps of the X-11 monthly filter when the default option is considered are as follows.[4]

Stage 1:   Initial Estimates
(a) The initial trend estimate is formed by means of a centered thirteen-term MA filter:

$$y_t^{tc}(1) = SM(L)y_t = (1/24)(1 + L)(1 + L \cdots + L^{11})L^{-6}y_t.$$
(4.5)

(b) The initial "SI ratio."
The filtered series obtained in this step is a first estimate of the seasonal plus noise part of the series. For the various decompositions this seasonal plus noise ratio (SI ratio) is obtained as

$$y_t^{si}(1) = y_t/y_t^{tc}(1) \qquad \text{(M, PA)},$$
$$y_t^{si}(1) = y_t - y_t^{tc}(1) \qquad \text{(A)}.$$

(c) Preliminary estimate of seasonal factor.
A first estimate of the seasonal part is obtained by applying the filter:

$$\tilde{y}_t^s(1) = M_1(L)y_t^{si}(1) = (1/9)(L^S + 1 + L^{-S})^2 y_t^{si}(1), \qquad (4.6)$$

with $S = 12$. In order to have the seasonal components sum to unity over 1 year, the filter $SM(L)$ is applied once more. Doing this yields a first estimate of the seasonal factor, as below.

(d) Initial seasonal factor.
For the various decompositions, it is defined as

$$y_t^s(1) = \tilde{y}_t^s(1)/SM(L)\tilde{y}_t^s(1) \qquad \text{(M, PA)},$$
$$y_t^s(1) = \tilde{y}^s(1) - SM(L)\tilde{y}_t^s(1) \qquad \text{(A)}.$$

Steps (a) through (d) represent the first stage and yield a first estimate of the seasonal component.

(e) Initial seasonal adjustment:

$$y_t^{sa}(1) = y_t/y_t^s(1) \qquad\qquad\qquad \text{(M)},$$
$$y_t^{sa}(1) = y_t - y_t^s(1) \qquad\qquad\qquad \text{(A)},$$
$$y_t^{sa}(1) = y_t - y_t^{tc}(1)(y_t^s(1) - 1) \qquad \text{(PA)}.$$

---

[4] Default option means that no extreme value corrections are performed (see Section 4.3.2) and that sufficient past and future values are available to make forecasting and backcasting unnecessary. The discussion in this section follows closely Ghysels and Perron (1993) and Findley et al. (1998, Appendix A).

Stage 2: Seasonal Factors and Seasonal Adjustment

The first operation of the second stage is to detrend the initially adjusted series by a $(2H + 1)$-term Henderson moving average filter. The selection of H is itself a delicate process described in detail by Findley et al. (1998, Appendix B). The default value is $H = 6$, yielding a thirteen-term Henderson moving average filter.

(a) Intermediate trend.

With the default Henderson filter, this is

$$y_t^{tc}(2) = HM(L) y_t^{sa}(1) = [-0.019L^6 - 0.028L^5 + 0.066L^3$$
$$+ 0.147L^2 + 0.214L + 0.240 + 0.214L^{-1} + 0.147L^{-2}$$
$$+ 0.066L^{-3} - 0.028L^{-5} - 0.019L^{-6}] y_t^{sa}(1). \quad (4.7)$$

The remainder of Stage 2 is concerned with refining the seasonal adjustment.

(b) Second SI ratio.

$$y_t^{si}(2) = y_t/y_t^{tc}(2) \qquad \text{(M, PA)},$$
$$y_t^{si}(2) = y_t - y_t^{tc}(2) \qquad \text{(A)}.$$

(c) Second seasonal factor by means of "3 × 5" seasonal moving average filter:

$$\tilde{y}_t^s(2) = M_2(L)y_t^{si}(2) = (1/15)\left(\sum_{j=-1}^{1} L^{jS}\right)\left(\sum_{j=-2}^{2} L^{jS}\right)y_t^{si}(2),$$
$$(4.8)$$

again with $S = 12$. Finally, in order to have the seasonal components sum to unity, the filter $SM(L)$ is applied again.

(d) Second seasonal factor.

$$y_t^s(2) = \tilde{y}_t^s(2)/SM(L)\tilde{y}_t^s(2) \qquad \text{(M, PA)},$$
$$y_t^s(2) = \tilde{y}_t^s(2) - SM(L)\tilde{y}_t^s(2) \qquad \text{(A)}.$$

(e) Second seasonal adjustment.

$$y_t^{sa}(2) = y_t/y_t^s(2) \qquad \text{(M)},$$
$$y_t^{sa}(2) = y_t - y_t^s(2) \qquad \text{(A)},$$
$$y_t^{sa}(2) = y_t - y_t^{tc}(2)(y_t^s(2) - 1) \qquad \text{(PA)}.$$

Stage 3: Final Henderson Trend and Final Irregular

The seasonally adjusted values from Stage 2 are used to produce final estimates of the trend and irregular components.

(a) Final trend.

We need again to select parameter H (not necessarily coinciding with the previously selected value) and compute (assuming here again the default value $H = 6$)

$$y_t^{tc}(3) = HM(L)y_t^{sa}(2).$$

(b) Final irregular.

$$y_t^i(3) = y_t^{sa}(2)/y_t^{tc}(3) \qquad \text{(M, PA)},$$
$$y_t^i(3) = y_t^{sa}(2) - y_t^{tc}(3) \qquad \text{(A)}.$$

This yields the final estimated decomposition:

$$y_t = y_t^{tc}(3)y_t^s(2)y_t^i(3) \qquad \text{(M)},$$
$$y_t = y_t^{tc}(3) + y_t^s(2) + y_t^i(3) \qquad \text{(A)},$$
$$y_t = y_t^{tc}(2)(y_t^s(2) - 1) + y_t^{tc}(3)y_t^i(3) \qquad \text{(PA)}.$$

For the additive decomposition, the three stages put together yield $v_{X-11}^M(L)$, namely the linear approximation to the monthly X-11 filter:

$$\begin{aligned}
v_{X-11}^M(L) &= 1 - SM_C(L)M_2(L)\{1 - HM(L) \\
&\quad \times [1 - SM_C(L)M_1(L)SM_C(L)]\} \\
&= 1 - SM_C(L)M_2(L) + SM_C(L)M_2(L)HM(L) \\
&\quad - SM_C^3(L)M_1(L)M_2(L)HM(L) \\
&\quad + SM_C^3(L)M_1(L)M_2(L), \qquad (4.9)
\end{aligned}$$

where $SM_C(L) = 1 - SM(L)$. The coefficients of this filter are presented in Table 4.1. Each coefficient is applied twice, once for the lag and once for the lead terms. The final two-sided symmetric monthly X-11 filter, appearing in (4.9), will be used to study the effects of filtering on tests for unit roots in Chapter 5. The Henderson MA filter is important, although it is only a subfilter, because it represents the trend estimate used in the program.[5] The transfer functions of both filters are reported in Figures 4.1 and 4.2. The transfer function of the final X-11 monthly filter has a large dip at the seasonal frequency and wiggles around one at all other frequencies. The Henderson MA filter of (4.7) is essentially a smoothing filter with a transfer function that equals one at low frequencies and drops off to zero at high frequencies.

The quarterly X-11 filter, analyzed by Laroque (1977), is constructed in a similar fashion. The final filter denoted by $v_{X-11}^Q(L)$ is composed of

$$SQ(L) = (1/8)(1 + L)(1 + L + L^2 + L^3)L^{-2}$$
$$HQ(L) = -0.073L^2 + 0.294L + 0.558$$
$$+ 0.294L^{-1} - 0.073L^{-2}, \qquad (4.10)$$

---

[5] The Australian Bureau of Statistics publishes Henderson trend estimates, based on the argument that they are less volatile than seasonally adjusted data. See Findley et al. (1998), p. 137 for further discussion.

Table 4.1. *Filter weights of the linear monthly X-11 filter*

Lags and leads						
0	0.819	23	0.011	46	−0.003
1	0.019	24	−0.121	47	−0.005
2	0.018	25	0.013	48	−0.005
3	0.017	26	0.013	49	−0.003
4	0.016	27	0.013	50	−0.001
5	0.015	28	0.013	51	0.002
6	0.014	29	0.012	52	0.003
7	0.013	30	0.008	53	0.003
8	0.014	31	0.005	54	0.002
9	0.015	32	0.004	55	0.001
10	0.018	33	0.003	56	0.000
11	0.020	34	0.003	57	−0.001
12	−0.179	35	0.004	58	−0.001
13	0.021	36	−0.063	59	−0.001
14	0.020	37	0.005	60	−0.001
15	0.018	38	0.007	61	−0.001
16	0.016	39	0.008	62	−0.001
17	0.015	40	0.008	63	0.001
18	0.012	41	0.008	64	0.001
19	0.009	42	0.005	65	0.001
20	0.009	43	0.002	66	0.000
21	0.009	44	0.001	67	0.000
22	0.010	45	−0.001	68	0.000

Source: Ghysels (1984, Table A.3.3.).

for initial trend estimation and the Henderson filter, along with $M_1(L)$ and $M_2(L)$, evaluated at $S = 4$. Again using (4.9), one can calculate the implied weights for the filter. They appear in Table 4.2. The transfer functions of the quarterly final filter and its Henderson MA subfilter are reported in Figures 4.3 and 4.4, respectively. They exhibit features similar to the monthly filters.

Bell (1992) provides a thorough analysis of the factorization of the linear X-11 filter polynomial. In particular, he shows that the linear X-11 contains the $(1 + L + \cdots + L^{S-1})$ polynomial but *not* the $(1 + L^{-1} + \cdots + L^{-S+1})$ polynomial. He also shows that the Henderson filter with default value $H = 6$ contains $(1 - L)^3(1 - L^{-1})^3$, which means the filter annihilates a polynomial trend up to degree 5.

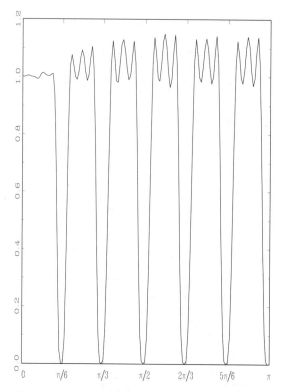

Figure 4.1. Transfer function of the linear monthly X-11 filter.

4.3.2   On Potential Sources of Nonlinearity

The question whether seasonal adjustment procedures are, at least approximately, linear data transformations is essential for several reasons. For instance, much of what is known about seasonal adjustment and estimation of regression models, which is the subject of Chapter 5, rests on the assumption that the process of removing seasonality can be adequately presented as a linear (two-sided and symmetric) filter applied to the raw data. Moreover, the theoretical discussions regarding seasonal adjustment often revolve around a linear representation (see also the discussion on the TRAMO/SEATS program Chapter 4, in Section 4.5).

Young (1968) investigated the question whether the linear filter was an adequate approximation and found it to be a reasonable proxy to the operations of the actual program. This result was, to a certain extent, a basic motivation as to why the linear filter representation has been extensively used in the literature. Sims (1974) and Wallis (1974), for instance, refer to Young's work to justify

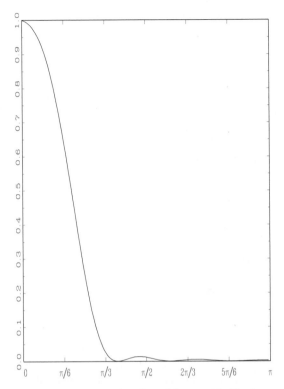

Figure 4.2. Transfer function of the monthly Henderson ($H = 6$) MA filter.

their analyses of linear filtering in linear regression models. Ghysels, Granger, and Siklos (1996) re-examined the question posed by Young. They came to quite the opposite conclusion, namely that the standard seasonal adjustment procedure is far from being a linear data-filtering process. They reached this different conclusion primarily because they took advantage of several advances in the analyses of time series, developed since Young (1968), and the leaps in the computational power of computers, which enabled them to conduct simulations that could not easily have been implemented before.

We will identify in this section the features contained in the X-11 program that may be sources of nonlinearity. Since the program is most often applied to monthly data, we cover this case exclusively and ignore the quarterly program.

As noted in Section 4.2, one must distinguish between different decompositions: additive, multiplicative, log-additive, and pseudo-additive. In principle only the additive and log-additive can be considered as linear filters, and then only if the default options are considered. Hence, neither the multiplicative nor the pseudo-additive filters are linear. Two features, specific to the multiplicative

Table 4.2. *Filter weights of the linear quarterly X-11 filter*

Lags and leads			
0	0.856	14	0.016
1	0.051	15	−0.005
2	0.041	16	−0.010
3	0.050	17	0.000
4	−0.140	18	0.008
5	0.055	19	−0.002
6	0.034	20	−0.003
7	0.029	21	0.000
8	−0.097	22	0.002
9	0.038	23	0.000
10	0.025	24	0.000
11	0.012	25	0.000
12	−0.053	26	0.000
13	0.021	27	0.000

Source: Laroque (1977, Table 1).

version, are also potential sources of significant nonlinearity. First, despite the multiplicative structure of the decomposition in (4.2), the program equates the 12-month *sums* of the seasonally adjusted and unadjusted data rather than their products. Second, the multiplicative version of the X-11 program also involves arithmetic means rather than geometric mean operations. We now turn to more general issues related to potential nonlinearity.

Multiplicative versus Additive

The bulk of economic time series handled by the U.S. Bureau of the Census and the U.S. Bureau of Labor Statistics are adjusted with the multiplicative version of the program. Only a small portion is treated with the additive version, apparently around 1% of the 3,000 series covered by the two aforementioned agencies. The Federal Reserve uses the additive version more frequently, because of the nature of the time series it treats.

Outlier Detections

The treatment of extremes, or outliers, is a key element in seasonal adjustment programs such as X-11 and even more in X-12-ARIMA, which is discussed in the next section. Because this feature is fundamentally the same for the additive and multiplicative versions, we will discuss it by using the former as an example.

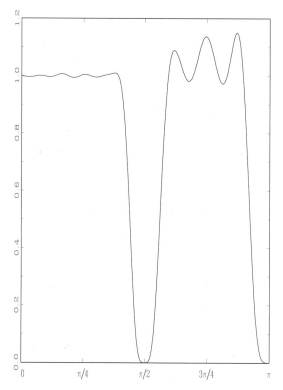

Figure 4.3. Transfer function of the linear quarterly X-11 filter.

We noted in the previous subsection that the X-11 program produces a first estimate of the seasonal and irregular components $y_t^{si}(1)$ and the seasonal factors $y_t^s(1)$. This yields a first estimate of the irregular, in the additive case equal to $y_t^i(1) = y_t^{si}(1) - y_t^s(1)$. The scheme to detect outliers is activated at this stage. First, a moving 5-year standard deviation of the estimated $y_t^i(1)$ process is computed. Hence, extractions of $y_t^i(1)$ will be evaluated against a standard-error estimate only involving the surrounding 5 years, that is, sixty observations in a monthly setting. We shall denote the standard error applicable to $y_t^i(1)$ as $\sigma_t^{(1)}$, where the superscript indicates that one has obtained a first estimate. The standard error is re-estimated after removing any observations on $y_t^i(1)$ such that $|y_t^i(1)| > 2.5\,\sigma_t^{(1)}$, yielding a second estimate $\sigma_t^{(2)}$, where the number of observations entering the second estimate is random because of the removal of observations. The second-round estimated standard error $\sigma_t^{(2)}$ is used to clear the $y_t^i(1)$ process from outlier or influential observations. The rules followed to purge the process can be described as follows.

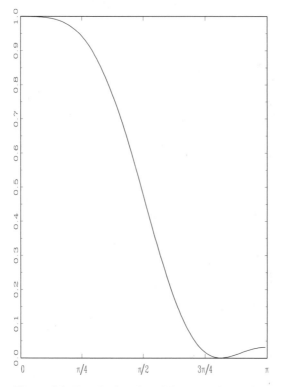

Figure 4.4. Transfer function of the quarterly Henderson ($H = 2$) MA filter.

1. A weighting function w_t is defined as

$$w_t = \begin{cases} 1 & \text{if } 0 \leq \left|y_t^i(1)\right| \leq 1.5\ \sigma_t^{(2)} \\ 2.5 - \left|y_t^i(1)\right|/\sigma_t^{(2)} & \text{if } 1.5\ \sigma_t^{(2)} < \left|y_t^i(1)\right| \leq 2.5\ \sigma_t^{(2)}. \\ 0 & \left|y_t^i(1)\right| > 2.5\ \sigma_t^{(2)} \end{cases}$$

(4.11)

2. $y_t^{si}(1)$ is replaced by nearest neighbor weighted average if

$$w_t < 1.$$

(4.12)

The criterion in (4.12) is used to replace any perceived outlier by the smoothed nearest neighbor estimate. The 1.5 and 2.5 values in (4.11), setting the benchmarks of the weighting function, play of course a key role besides the two-step standard-error estimate $\sigma_t^{(2)}$. The (4.11)–(4.12) scheme is, however, entirely based on rules of thumb and not so easy to rationalize. The value of

$1.5\,\sigma_t^{(2)}$ in (4.11) that sets off the correction scheme, since it determines whether $w_t < 1$, is quite tight.

Moving Average Filter Selection

We will continue with the additive version of the program, again for the sake of discussion. The X-11 procedure continues with the second stage, with as a first step the extraction of the trend-cycle component by applying a Henderson moving average filter. The selection of H may differ from its default value $H = 6$ considered in the previous subsection. In practice the value of H can change from one period to the next. To describe the scheme, let us define two average percentage changes: μ_{1t} is the average absolute month-to-month change of $y_t^{sa}(1) - y_t^{tc}(1)$, and μ_{2t} is the average absolute month-to-month change of $y_t^{tc}(1)$. The averages are updated as new raw data are added to the sample and are therefore made time varying. The filter selection scheme can then be formulated as follows.

1. Apply nine-term Henderson MA $(H = 4)$ if

$$\mu_{1t} < 0.99\,\mu_{2t}. \tag{4.13}$$

2. Apply thirteen-term Henderson MA $(H = 6)$ if

$$0.99\,\mu_{2t} \le \mu_{1t} < 3.5\,\mu_{2t}. \tag{4.14}$$

3. Apply twenty-three-term Henderson MA $(H = 11)$ if

$$3.5\,\mu_{2t} \le \mu_{1t}. \tag{4.15}$$

This selection procedure is repeated at the third stage when the Henderson filter is reapplied (in the previous subsection H was set twice to its default value). The X-12-ARIMA program has more options on the choice of H. This will be discussed in the next section.

Aggregation

So far, we have highlighted features that represent the possible causes of nonlinearity and/or time variation in the actual X-11 filtering process. However, another source of nonlinearity also has to be mentioned. It is not related to the intrinsic operational rules of the program but rather to the modus operandi of its application to several series.

Seasonal adjustment procedures are quite often applied to disaggregated series, like narrowly defined industrial sectors or components of monetary aggregates, and the output is then aggregated to produce a seasonally adjusted aggregate. Obviously, the separate decomposition (4.1) for two series, say x_t and y_t, is not the same as the decomposition for a z_t process defined as $z_t \equiv x_t + y_t$. The question whether seasonal adjustment should precede or follow aggregation is discussed in Geweke (1978) and was recently re-examined by Ghysels (1997a) and will be discussed in Section 4.6. When the seasonal adjustment process is linear and uniform, then aggregation and seasonal adjustments are interchangeable. Another potential source of nonlinearity is introduced, however, when seasonal adjustment and aggregation are not interchangeable, and one applies the procedure to disaggregated series with only the aggregated series available to the public. In practice, this setup is quite common. In particular, seasonally adjusted aggregate industrial production, inventories, consumption, and shipments series are usually obtained by means of aggregation of adjusted component series (like aggregation across industries, etc.). One of the consequences of this common practice is that aggregate adjusted series are not the seasonally adjusted version of the aggregate raw series, since seasonal adjustment procedures are applied at the disaggregate level.

4.4   The X-12 Seasonal Adjustment Program

The X-11 procedure underwent several improvements, some of which were major changes to the program. One major improvement was designed and implemented by Statistics Canada and called X-11-ARIMA [Dagum (1980)] that had the ability to extend time series with forecasts and backcasts from ARIMA models prior to seasonal adjustment. Its main advantage was smaller revisions of seasonally adjusted series as future data became available [see, e.g., Bobbitt and Otto (1990)]. The X-11-ARIMA program also featured improved diagnostics used to appraise the quality of adjusted series. Such diagnostics are designed to assess whether there are seasonal patterns unaccounted for or whether seasonal patterns change through time.

The U.S. Census Bureau proceeded in 1998 to major improvements of the X-11 procedure as well. These changes were so important that they prompted the release of what is called X-12-ARIMA. Findley et al. (1998) provide a very detailed description of the new improved capabilities of the X-12-ARIMA procedure. It encompasses the improvements of Statistics Canada's X-11-ARIMA and encapsules it with a front end regARIMA program, which handles regression and ARIMA models, and a set of diagnostics, which enhance the appraisal of the output from the original X-11-ARIMA. Hence, the X-11-ARIMA program is still the core of the procedure.[6] Before series are fed into X-12-ARIMA, they are preadjusted by the regARIMA routine, a procedure that pre-empts the input series of features that are known to be inadequately dealt with by the original X-11-ARIMA program. In the first subsection, we discuss the regARIMA part of the X-12-ARIMA procedure; the second subsection covers the postadjustment diagnostics.

---

[6] The X-12 program adds some longer seasonal moving averages and Henderson trend filters.

4.4.1 RegARIMA Modeling in X-12-ARIMA

The procedure considers models of the following type:

$$\phi_p(L)\phi_P(L^S)(1-L)^d(1-L^S)^D\left(y_t - \sum_{i=1}^{r}\beta_i x_{it}\right) = \theta_q(L)\theta_Q(L^S)\varepsilon_t, \quad (4.16)$$

where $\phi_p(z)$, $\phi_P(z)$, $\theta_q(z)$, and $\theta_Q(z)$ are polynomials of degree p, P, q, and Q respectively. Moreover, the input series y_t is typically obtained from a nonlinear transformation of the original series.[7] The model in (4.16) can also be written as

$$(1-L)^d(1-L^S)^D y_t = \sum_{i=1}^{r}\beta_i\{(1-L)^d(1-L^S)^D x_{it}\} + \omega_t,$$

where $\omega_t = (1-L)^d(1-L^S)^D(y_t - \sum_{i=1}^{r}\beta_i x_{it})$, which is assumed to be a covariance stationary process. The regARIMA program has a set of built-in regressors for the monthly case [listed in Table 2 of Findley et al. (1998)]. They include a constant trend, deterministic seasonal effects, trading-day effects (for both stock and flow variables), length-of-month variables, leap year, Easter holiday, Labor day, and Thanksgiving dummy variables as well as additive outlier (AO), level shift (LS), and temporary ramp regressors.

The preadjustment regARIMA procedure therefore contains many operations that are also featured in the original X-11 program. For instance, before series are fed into X-11, they can be treated by an automatic outlier treatment for AO and LS by means of the regARIMA procedure.[8] Alternatively, the regARIMA preadjustment for outliers could be ignored and the original X-11 outlier procedure described in Subsection 4.3.2 could be applied. Likewise, calendar and holiday adjustments can be performed with the regARIMA program either before applying X-11 or else in the original program known as the Young (1965) method. One might expect that estimates of calendar and outlier effects are better treated by the preadjustment regARIMA program, because as Findley et al. (1998) note, the regression models account for the correlation structure of the observed series and treat calendar and outlier effects directly instead of at the level of the estimated irregular $y_i^I(1)$ component. Unfortunately, as Findley et al. (1998) also note, this is not universally true; that is, sometimes not using the regARIMA preadjustment option and relying entirely on the X-11

---

[7] The X-12-ARIMA procedure allows for Box–Cox transformations; see Findley et al. (1998) Equation (9) for more detail. In principle one can also consider $y_t = \log Y_t - \log d_t$, where d_t is a divisor that may include deseasonalization, detrending, or length-of-month factors. Obviously, it is not clear whether such preadjustment is desirable since the X-12 program is built for the purpose of adjustment.

[8] The stepwise regression-based automatic outlier procedures in regARIMA are based on the work of Chang and Tiao (1983).

program is better. They provide an example and suggest the use of postadjust-ment diagnostics to decide which approach to take.

4.4.2   X-12 Diagnostics

The purpose here is only to provide a brief description of the diagnostics used to appraise the quality of X-11 adjustment output. The postadjustment diagnos-tic tables of X-11 and X-11-ARIMA were incorporated in X-12-ARIMA. In addition, new diagnostics were added. First, X-12-ARIMA produces estimates of two spectra, namely the spectrum of (1) the month-to-month differences of the adjusted series (or log of the series in the case of a multiplicative decom-position) and of (2) the final irregular component y_t^i (3) adjusted for extreme values by using the X-11 outlier detection procedure. The program inspects in particular the trading-day sensitive frequencies of 0.304, 0.348, and 0.432 cycles/month (plotted with a T symbol), as well as the seasonal frequencies $k/12$ for $k = 1, \ldots, 6$ (plotted with an S symbol). Visual inspection of peaks is a key ingredient here in the diagnostic procedure. Last but not least, the stability of seasonal patterns is investigated. It is widely known that seasonals tend to vary through time [see, e.g., Canova and Ghysels (1994)]. The use of sliding spans of data, a technique proposed by Findley et al. (1990), is implemented in the X-12-ARIMA procedure. Stability is also investigated by means of so-called revision histories, that is, the process of updating seasonal factors for any given time t as new (future) data becomes available.

4.5   The TRAMO/SEATS Programs

There is a long tradition of viewing seasonal adjustment as a signal extraction problem. In particular, for more than three decades now, seasonal adjustment has been portrayed in the context of spectral domain representations. See, for instance, Hannan (1963), Granger and Hatanaka (1964), Nerlove (1964), and Godfrey and Karreman (1967), among others. The frequency-domain analysis led to the formulation of seasonal adjustment as a signal extraction problem in a linear unobserved component ARIMA (henceforth UCARIMA) framework, where the optimal minimum mean-squared error filters are linear.

The theory of signal extraction involving nonstationary processes, which will be the case covered here, was developed by Hannan (1967), Sobel (1967), Cleveland and Tiao (1976), Pierce (1979), Bell (1984), Burridge and Wallis (1988), and Maravall (1988). While the theory is well developed, the prac-tical implementation remained for a long time very challenging. Goméz and Maravall (1996) succeeded in building a seasonal adjustment package using signal extraction principles. The package consists of two programs, namely TRAMO (Time Series Regression with ARIMA Noise, Missing observations, and Outliers) and SEATS (Signal Extraction in ARIMA Time Series). The TRAMO program fulfills the role of preadjustment, very much like regARIMA does for X-12-ARIMA adjustment. Hence, it performs adjustments for outliers, trading-day effects, and other types of intervention analysis [following Box and Tiao (1975)]. We will not cover the TRAMO program in any further detail because of its resemblance to the regARIMA procedure covered earlier.[9]

The use of signal extraction principles requires the specification of signal and noise components. This approach is often called model-based seasonal adjustment because each of the components in, say, an additive decomposition requires an explicit time series model specification. ARIMA models are used to describe each of the components; hence the term UCARIMA. The SEATS program is precisely built to extract seasonal components from observed time series by using an UCARIMA structure. In the first subsection, we describe the basic structure of model-based seasonal adjustment as it is implemented in the SEATS program. In the second subsection, we compare this approach with that of X-12.

4.5.1 Unobserved Component ARIMA Models and Seasonal Adjustment

For simplicity, let us consider an additive decomposition, namely

$$y_t = y_t^{tc} + y_t^s + y_t^i,$$

which corresponds to (4.4) previously discussed. For the purpose of presentation it will be convenient to introduce the nonseasonal component that is defined as $y_t^{ns} = y_t^{tc} + y_t^i$ and hence $y_t = y_t^{ns} + y_t^s$. Unlike the setup of the linear X-11 program, we start with specifying parametric linear stochastic ARIMA processes for each of the components. Since none of the components is observed, we obviously also need to impose identification assumptions. Bell and Hillmer (1984) and Goméz and Maravall (1996), among many others, describe such assumptions, which can be summarized as follows.[10]

UCARIMA Model Assumptions: It is important to note here that we explicitly define time series models for the components y_t^{ns} and y_t^s. Therefore, we specify in this section all polynomials pertaining to the former component with the index ns, whereas the index s applies to the seasonal component.

A.1: $\{y_t^{ns}\}$ and $\{y_t^s\}$ are mutually independent.

A.2: $\{y_t^{ns}\}$ follows an ARIMA model $\phi_{ns}(L)\Delta_{ns}(L)y_t^{ns} = \theta_{ns}(L)u_t^{ns}$ where u_t^{ns} is i.i.d. $N(0, \sigma_{ns}^2)$.

---

[9] This does not mean that certain features do not differ between regARIMA and TRAMO. Goméz and Maravall (1996) provide a User's Guide, which contains the specific details for their program.

[10] We assume absence of deterministic seasonal components.

We describe later how the polynomials of the components here relate to the ARIMA representation for the observed series, similar to, for instance, (4.16) for X-12-ARIMA. The components are assumed to satisfy:

A.3: $\{y_t^s\}$ follows an ARIMA model $\phi_s(L)\Delta_{(s)}(L)y_t^s = \theta_s(L)u_t^s$, where u_t^s i.i.d. $N(0, \sigma_s^2)$.

A.4: The polynomials $\phi_{ns}(L)$ and $\phi_s(L)$ have no common zeros.

A.5: The polynomials $\Delta_{ns}(L)$ and $\Delta_{(s)}(L)$ share no roots on the unit circle and $\Delta_{(s)}(L) = (1 + L + \cdots + L^{S-1})$.

A.6: $\theta_s(L)$ is of order less than or equal to $S - 1$.

A.7: The variance of u_t^s is minimal among all decompositions satisfying A.1 through A.6.

These identification assumptions are obviously not harmless, and indeed much has been written about the orthogonality assumption A.1, which is not theory free, and the theory behind it rarely corresponds to standard economic theory developments. This is forcefully argued in Ghysels (1988, 1994a). Assumptions A.2–A.6 reflect the notion that the seasonal component captures the spectral peaks appearing at seasonal frequencies. Indeed, the annual summation operator $\Delta_{(s)}(L)$ of Assumption A.5 is extensively discussed in Chapter 3 as implying the presence of unit roots at all seasonal frequencies. The last assumption, A.7, deals with the non-uniqueness of orthogonal decompositions into seasonal and nonseasonal components. This assumption is not always imposed, as will be discussed later.

Following Assumption A.5, we know that $\Delta_{ns}(L) = (1 - L)$, or else $\Delta_{ns}(L) = (1 - L)^2$. Then, as shown by Pierce (1979) and Bell (1984), among others, the time series properties of the process $y_t = y_t^{ns} + y_t^s$ has the following structure: $\phi(L)\Delta_y(L)y_t = \theta(L)u_t$, where (1) $\Delta_y(L) \equiv \Delta_{ns}(L)\Delta_{(s)}(L)$, (2) u_t is a white-noise process with variance σ^2, and (3) the polynomials $\phi(L)$ and $\theta(L)$ can be determined from the covariance generating function, namely

$$\sigma^2\phi(L^{-1})\theta(L^{-1})\phi(L)\theta(L)$$
$$= \sigma_{ns}^2\Delta_{(s)}(L^{-1})\Delta_{(s)}(L)\phi_{ns}(L)\theta_{ns}(L)\phi_{ns}(L^{-1})\theta_{ns}(L^{-1})$$
$$+ \sigma_s^2\Delta_{ns}(L^{-1})\Delta_{ns}(L)\phi_s(L)\theta_s(L)\phi_s(L^{-1})\theta_s(L^{-1}). \quad (4.17)$$

The time series model $\phi(L)\Delta_y(L)y_t = \theta(L)u_t$ involving the polynomials specified above can be compared with the model appearing in (4.16) on which the X-12-ARIMA program is built. It is clear that the model specification is quite different since in (4.16) one has potentially a multiplicative model involving the unit roots $(1 - L)(1 - L^S)$. Moreover, the polynomials $\phi(L)$ and $\theta(L)$ are constrained as they are tied to the component model specifications. This suggests at least two issues with regard to model specification. First and foremost, it should be useful to conduct seasonal unit root tests to determine the order of integration of the unadjusted series y_t. Second, it is also important to determine properly the lag order of the polynomials specifying the component models, that is, $\phi_{ns}(L)$, $\phi_s(L)$, $\theta_{ns}(L)$, and $\theta_s(L)$. We will discuss this issue in the next subsection. We first cover the estimation of the component model.

There are various ways to present the theory of optimal signal extraction for linear time series processes. One way involves the use of state-space models and Kalman filtering; see for instance, Harvey (1989) and Goméz and Maravall (1994), among others, for applications to seasonal processes. Another approach to linear signal extraction theory relies on frequency-domain analysis. The second approach will be used in the next and last section of the chapter. Here, we will use a state-space model approach.

Suppose that the order of the $\phi_{ns}(L)$ polynomial is p_{ns} and that of the $\theta_{ns}(L)$ polynomial is q_{ns}. Likewise, suppose $\phi_s(L)$ is of order p_s and $\theta_s(L)$ is of order q_s. Then, define $r_j = \max(p_j, q_j)$ for $j = ns, s$. We construct the following linear systems for y_t^{ns} and y_t^s:

$$X_t^j = F_j X_{t-1}^j + G_j u_t^j, \qquad j = ns, s,$$
$$y_t^j = H_j' X_t^j, \qquad\qquad j = ns, s,$$

where X_t^j is the $r_j \times 1$ state vector with the first element $X_t^j(1)$ equal to y_t^j and the other entries to the vector:

$$X_t^j(k) = \sum_{i=k}^{r_j} \phi_{ji} y_{t-1+k-i}^j - \sum_{i=k-1}^{q_j} \theta_{ji} u_{t-1+k-i}^j,$$

for $k = 2, \ldots, r_j$. The vector $H_j' = (1, 0, \ldots, 0)'$ is of dimension r_j and $G_j = (1, \theta_{j1}, \ldots, \theta_{jr_j})'$ is a $r_j \times 1$ vector with $\theta_{ji} = 0$ for $i > q_j$. Finally, the matrix F_j is $r_j \times r_j$ and has the following structure:

$$F_j = \begin{pmatrix} \phi_{j1} & 1 & 0 \cdots 0 \\ & & 1 \\ \vdots & \vdots & \ddots \\ & 0 & & 1 \\ \phi_{jr_j} & 0 & & 0 \end{pmatrix},$$

where $\phi_{ji} = 0$ for $i > p_j$. Then the state-space representation of the component model is of degree $r = r_{ns} + r_s$ and

$$y_t = H' X_t + y_t^i,$$
$$X_t = F X_{t-1} + G V_t,$$

where the $r \times 1$ state vector is $X_t' = [(X_t^{ns})'(X_t^s)']$; $H' = (H_{ns}', H_s')$; $F = \text{diag}(F_{ns}, F_s)$; $G = \text{diag}(G_{ns}, G_s)$; and $V_t' = [u_t^{ns}, u_t^s]$. The innovations u_t^j are typically assumed to be $N(0, c_j \sigma_i^2)$ where σ_i^2 is the variance of the irregular component y_t^i i.i.d. $N(0, \sigma_i^2)$. Hence, c_{ns} is the trend-noise innovation ratio and c_s is the seasonal-noise innovation ratio. One sets $\sigma_i^2 = 1$ to resolve the non-uniqueness problem.[11]

The Kalman filter recursions, conditional on initial conditions $X_{1|0}$ and $P_{0|0}$, are

$$X_{t|t-1} = F X_{t-1|t-1}, \qquad P_{t|t-1} = F P_{t-1|t-1} F' + GG',$$
$$X_{t|t} = X_{t|t-1} + K_t(y_t - H'X_{t|t-1}),$$
$$P_{t|t} = P_{t|t-1} - K_t H' P_{t|t-1},$$

where $K_t = P_{t|t-1} H (H' P_{t|t-1} H + 1)^{-1}$; $X_{t|t-1} = E[X_t | y_{t-1}, y_{t-2}, \ldots]$; and $P_{t|t-1} = \text{Var}[X_t | y_{t-1}, y_{t-2}, \ldots]$. Likewise, the vector $X_{t|t}$ contains the concurrent extraction of the unobserved components y_t^{ns} and y_t^s, given data on $\{y_t, y_{t-1}, \ldots\}$. Moreover, $P_{t|t}$ provides a conditional variance of the components.[12] The Kalman filter setup also readily yields a maximum likelihood procedure for parameter estimation. In particular, let $v_t = y_t - H'X_{t|t-1}$ and $\sum_t = H' P_{t|t-1} H + 1$. Then the likelihood function can be written as

$$\ell = -\frac{1}{2} \left[\ell n \left(\prod_{t=1}^{T} \sum_t \right) + T \, \ell n(e'e) \right],$$

where T is the sample size while $e = (e_1, \ldots, e_T)'$ with $e_t = v_t / \Sigma_t$. Moreover, because $\sigma_i^2 = 1$, the parameter σ_i^2 has been concentrated out of the likelihood function. Seasonal adjustment applications typically involve nonstationary time series. The initialization of the Kalman filter is therefore nontrivial as there is no marginal distribution. Several solutions have been suggested; for further discussion see Bell (1984), Harvey and Pierse (1984), Kohn and Ansley (1986), de Jong (1991), de Jong and Lin (1994), or Goméz and Maravall (1994). Goméz (1999) establishes the equivalence of three of these approaches.

4.5.2   Model-Based Approach versus X-12

The brief discussion in the previous section regarding (4.16) and (4.17) hinted at the fact that there is a fundamental difference between the TRAMO/SEATS program, and in general the so-called model-based approach, and the X-12-ARIMA routines. The former designs a filter tailored to each series, whereas the latter applies, at least under ideal circumstances, a uniform filter across all

---

[11] This circumvents imposing Assumption A.7.

[12] The X-11 and related programs do not provide measures of precision and this omission has been criticized; see, e.g., Burridge and Wallis (1984) for further discussion. For recent attempts to include standard errors for X-11 and X-12/ARIMA, see, e.g., Pfeffermann (1994), Pfeffermann, Morry, and Wong (1995), and Bell and Kramer (1999).

series regardless of the time series structure of the seasonal and nonseasonal components.[13] This fundamental difference has far-reaching consequences. Some of the consequences will be discussed in the next section; others will be dealt with in Chapter 5, where estimation and filtering issues will be covered. Before addressing these fundamental issues, let us first elaborate further on the comparisons of the two approaches.

In Subsection 4.3.1, we discussed the linear X-11 filter. From the previous section, we also know that optimal signal extraction filters are also linear. One may therefore wonder whether there is a particular UCARIMA model structure for which the X-11 filter is the optimal signal extraction filter. This question has been answered in the context of monthly data by Cleveland (1972). He found the following UCARIMA model:

$$(1 - L)^2 y_t^{ns} = (1 + 0.26L + 0.30L^2 - 0.32L^3)\varepsilon_t^{ns},$$

$$(1 + L + \cdots + L^{11})y_t^s = (1 + 0.26L^{12})\varepsilon_t^s,$$

yielding a time series model for the observed series of the following type:

$$(1 - L)(1 - L^{12})y_t = \theta(L)\varepsilon_t,$$

where $\theta(L)$ is a polynomial of order 14 involving specific coefficients [omitted here, but see Cleveland (1972)]. Hence, the linear X-11 filter happens to be optimal for a particular type of model; for any other time series process, it is therefore suboptimal in a linear optimal mean-squared error (MSE) signal extraction sense.[14] We also observe that X-11 involves no parameter estimation, in sharp contrast to model-based procedures. In practice, one may constrain the class of models from which to compute signal extraction filters to reduce the effect of parameter uncertainty. In some sense this is an intermediate solution between a general specification, as discussed in the previous section, and the linear X-11 program, which does not allow for any flexibility regarding filter weight design.[15]

The first of these more tightly parameterized model-based specifications is from Hillmer and Tiao (1982) and involves the following model structures:

$$(1 - L)^2 y_t^{ns} = (1 + \theta_{ns1}L)(1 + L)\varepsilon_t^{ns},$$

$$(1 + L + \cdots + L^{11})y_t^s = \theta_s(L)\varepsilon_t^s,$$

---

[13] By ideal circumstances we mean the absence of outliers, etc., which trigger off the intervention analysis features of the X-12-ARIMA program (as well as TRAMO/SEATS).

[14] Besides Cleveland (1972), see also Burridge and Wallis (1984) for further discussion on the optimal signal extraction interpretation of X-11.

[15] We do not consider here the fact that certain subfilters of X-11, such as the Henderson moving average filter, can be changed by selecting values for the parameter H. Moreover, H is not an estimated parameter in a typical sense of the word.

yielding a model for the observed time series:

$$(1 - L)(1 - L^{12})y_t = (1 + \theta_1 L)(1 + \theta_{12}L^{12})\varepsilon_t. \qquad (4.18)$$

This model involves only two parameters, obviously less than the general specification would call for.

Another structure, from Harvey and Todd (1983), implemented in the STAMP program [see Koopman et al. (1996)], selects the following setup:

$$(1 - L)^2 y_t^{ns} = (1 + \theta_1 L)\varepsilon_t^{ns},$$

$$(1 + L + \cdots + L^{11})y_t^s = \varepsilon_t^s,$$

yielding the process structure for y_t:

$$(1 - L)(1 - L^{12})y_t = \theta(L)\varepsilon_t, \qquad (4.19)$$

where $\theta(L)$ is a polynomial of order 13 involving θ_1 as the only free parameter. The discussion of the two specific models leads us to the subject of model selection.

The specification of UCARIMA models has followed two directions: (1) specify model structures for the component processes, that is, specify a structure for the $\phi_j(L)$ and $\theta_j(L)$ polynomials $j = ns, s$; or (2) characterize first the ARIMA models for the observed time series y_t and derive its components. The latter consists of fitting a tightly parameterized model such as (4.18) or (4.19) to the unadjusted data and inferring the implied component models. This strategy is implemented in the STAMP program. The TRAMO/SEATS program takes the more ambitious approach of allowing for a data-driven component model specification through (4.17).[16] Model selection in the automatic model identification setup of the TRAMO/SEATS program uses the Hannan and Rissanen (1982) method to select the order of the polynomials $\phi_j(L)$ and $\theta_j(L)$ for $j = ns$ and s and for hence specifying $\phi(L)$ and $\theta(L)$ in (4.17).

As just indicated, the model-based approach to seasonal adjustment requires the estimation of UCARIMA parameters for the specific series. Thus, in principle, the loss of the corresponding number of degrees of freedom should be taken into account when models are estimated by using the seasonally adjusted data. Clearly, this could have a nontrivial impact on the degrees of freedom used for inference. It is arguable that the use of a program such as X-12, which is based on a common filter, leads to a negligible loss of degrees of freedom because there is no estimation related to the specific series. This is, however, a naive view, because of all the adjustments built into X-12-ARIMA and outlined above. Therefore, the number of degrees of freedom lost is unclear.

---

[16] Maravall (1985) discusses the links between the two model specification approaches.

4.6 Seasonal Adjustment and Other Data Transformations

To conclude this chapter we consider an issue not often dealt with in discussions regarding seasonal adjustment. In applied research, seasonal adjustment is rarely the only data transformation. A prominent example of such a scenario is one in which economists study "real" or constant-dollar series obtained from deflating "nominal" raw data by price indexes. Constant-dollar data are also quite often divided by (seasonally adjusted) population series to yield a per capita measure. A common feature of this and numerous other examples is that seasonal adjustment *precedes* all other data transformations, as researchers collect seasonally adjusted series and then combine two or more such series ultimately to obtain the data of interest. The bulk of the literature on seasonal adjustment has focused on the design of procedures separate from other potential uses or operations with data such as estimation of regression models, temporal or cross-sectional aggregation, and so forth. The issue of seasonal adjustment and regression analysis will be dealt with in the next chapter. Ghysels (1997a) deals with seasonal adjustment and other data transformations that we cover in this section.

It can, *and quite likely will*, happen that adjustment by (ad hoc linear) X-11 is preferable to optimal signal-extraction individual series adjustment. That this *can* occur is not surprising. That this occurrence is likely *is* surprising and is the main finding of Ghysels (1997a). One can show that when the data transformations are done in the "wrong" order, applying a uniform seasonal adjustment procedure may easily yield *better* extractions of the series of interest to the final user. This has nonnegligible implications for adopting signal extraction as a principle of seasonal adjustment design. Indeed, when seasonal adjustment precedes all other transformations and univariate optimal signal-extraction filters are used, this often results in the most *inefficient* strategy in terms of MSE for the transformed series. Consequently, filter design along signal-extraction principles may not be as desirable as uniform filtering once the potential of other data transformations being present is acknowledged. Therefore, the case for uniform filters such as X-11 is much stronger than is commonly thought once it is recognized that seasonal adjustment is not the only transformation applied to data.

We will treat the issue of seasonal adjustment and other data transformations in its most general context. The theoretical results providing necessary and sufficient conditions to find out whether it is better for seasonal adjustment to precede other data transformations are discussed first. Various special cases, such as the uniform filter and optimal signal-extraction filter, are also considered. The formal framework also enables us to extend some of the results of Geweke (1978) on cross-sectional aggregation and seasonal adjustment.

The formal structure of the analysis can be described as follows: The series of interest is denoted y_t and its nonseasonal component is denoted y_t^{ns}. The series x_t and z_t are combined to yield y_t, by means of the relationship

$$x_t = y_t + z_t.$$

This relationship can be justified by the standard assumption that all processes are specified in logs. The idea implicitly behind this framework is that y_t and z_t are the primary behavioral variables of interest (like a price and quantity series) with x_t definitionally related to them (like a nominal series). Computationally, y_t is obtained from $x_t - z_t$. Though restricted to linear filtering, the analysis does not rely at first on any explicit form of filtering. It is assumed that $x_t^{ns} \equiv v_1(L)x_t$, $z_t^{ns} \equiv v_2(L)z_t$, and $x_t - z_t$ is filtered as $v_3(L)(x_t - z_t)$, where $v_i(L)$, $i = 1, 2, 3$ are any set of linear filters. The common practice is to calculate $y_t^{ns} = x_t^{ns} - z_t^{ns}$; that is, to apply other data transformations after seasonal adjustment.

There is a very significant and important difference between cases where $v_1(L) \equiv v_2(L) \equiv v_3(L)$ (that is, a uniform filter like X-11 is applied) versus cases where $v_i(L)$ differs across series, which occurs for instance when the linear filters are data based and their design is built on optimal signal-extraction principles. With a uniform filter, the order of seasonal adjustment and other transformations, provided they are linear, does not matter. When the filters differ, the order becomes quite critical. When seasonal adjustment is viewed as a signal-extraction problem, then it should most often come as the *last* transformation because the "signal" of interest is not the nonseasonal component of x_t or z_t but instead that of y_t. Obviously, optimal signal-extraction filters will always outperform (in terms of mean-squared extraction error for y_t^{ns}) uniform filters provided, however, that the transformations are done in the right order.

We must impose certain restrictions on the processes to have a well-defined seasonal adjustment problem. These restrictions are similar to those discussed in the previous section but now apply to a combination of two series. For convenience here we assume without loss of generality an MA representation. In particular:

Assumption A.8: The four components y_t^{ns}, y_t^s, z_t^{ns}, and z_t^s are described by the following processes:

$$\Delta^k y_t^{ns} \equiv (1-L)^k y_t^{ns} = \theta_{yns}(L)u_t^{yns} \qquad \text{for} \quad k = 1 \text{ or } 2, \qquad (4.20)$$

$$\Delta^{\tilde{k}} z_t^{ns} \equiv (1-L)^{\tilde{k}} z_t^{ns} = \theta_{zns}(L)u_t^{zns} \qquad \text{for} \quad \tilde{k} = 1 \text{ or } 2, \qquad (4.21)$$

$$\Delta_{(s)} y_t^s \equiv (1 + L + L^2 + \cdots L^{11})y_t^s = \theta_{ys}(L)u_t^{ys}, \qquad (4.22)$$

$$\Delta_{(s)} z_t^s \equiv (1 + L + L^2 + \cdots L^{11})z_t^s = \theta_{zs}(L)u_t^{zs}. \qquad (4.23)$$

The innovations have finite variances, respectively, equal to σ_{yns}^2, σ_{zns}^2, σ_{ys}^2, and σ_{zs}^2.

Note that the orders of differencing k and $\tilde{k}$ need not be equal so that the zero-frequency nonstationarity for the series of interest y_t and the component series x_t and z_t may differ. Such is the case, for instance, with nominal and real time series because nominal series (x_t) and prices (z_t) are often $I(2)$, whereas real series (y_t) are $I(1)$. Hence, our framework allows for cointegration between x_t and z_t, which cancels one of the zero-frequency unit roots. When $\tilde{k} = 2$ and $k = 1$, there is cointegration between x_t and z_t; otherwise $k = \tilde{k} = 1$ or 2. The most critical assumption, however, will be that the unobserved component series do not share common unit roots, as stated in Assumption A.8. This will guarantee that the signal-extraction filters and their MSE, to be defined, are well specified. In general, the nonseasonal growth rates of y_t and z_t will not necessarily be uncorrelated. Likewise, changes in the seasonal components of y_t and z_t may be correlated as well. These relationships are spelled out by the following assumption:

Assumption A.9: The processes $\Delta^{\tilde{k}} x_t^{ns}$, $\Delta_{(s)} x_t^{s}$, $\Delta^{\tilde{k}} z_t^{ns}$, $\Delta_{(s)} z_t^{s}$, $\Delta^{k} y_t^{ns}$, and $\Delta_{(s)} y_t^{s}$ satisfy

$$f_x^j(e^{-i\omega}) = f_y^j(e^{-i\omega}) + f_z^j(e^{-i\omega})$$

$$+ 2\,\text{Re}\big[f_{yz}^j(e^{-i\omega})\big] \quad -\pi \le \omega \le \pi, \tag{4.24}$$

for $j = ns$ and s, where $f_v^j(e^{-i\omega}) = \sigma_{vj}^2 |\theta_{vj}(e^{-i\omega})|^2$, $j = ns, s; v = z, y$; and $f_{yz}^{ns}(e^{-i\omega})$ represents the cospectrum between $\{\Delta^{\tilde{k}} z_t^{ns}\}$ and $\{\Delta^{k} y_t^{ns}\}$ while $f_{yz}^s(e^{-i\omega})$ is the cospectrum between $\{\Delta_{(s)} z_t^{s}\}$ and $\{\Delta_{(s)} y_t^{s}\}$.

Assumptions A.8 and A.9 and (4.17) yield the following stochastic structure for $\{y_t\}$:

$$(1 - L^{12})\Delta^{k-1} y_t = \theta_y(L) u_t^y, \qquad u_t^y \text{ iid } N(0, \sigma_y^2), \tag{4.25}$$

where $\Delta^{k-1} = 1$ for $k = 1$ and $\Delta^{k-1} = \Delta = (1 - L)$ for $k = 2$. A similar structure is obtained for x_t involving the one-sided polynomial $\theta_x(L)$ and error u_t^x i.i.d. $N(0, \sigma_x^2)$ and likewise for z_t involving $\theta_z(L)$ and error θ_t^z i.i.d. $N(0, \sigma_z^2)$.

Consider now two extraction regimes – one labeled "A," indicating that seasonal adjustment comes "after" the other data transformations, and the other labeled "B," corresponding to "before." The extraction regime B is defined as

$$\widehat{y}_{Bt}^{ns} = E\big[x_t^{ns}|\{x_t\}_{t=-\infty}^{+\infty}\big] - E\big[z_t^{ns}|\{z_t\}_{t=-\infty}^{+\infty}\big],$$

$$= E\big[y_t^{ns} + z_t^{ns}\big|\{y_t^{ns} + y_t^{s} + z_t^{ns} + z_t^{s}\}_{t=-\infty}^{+\infty}\big]$$

$$- E\big[z_t^{ns}\big|\{z_t^{ns} + z_t^{s}\}_{t=-\infty}^{+\infty}\big], \tag{4.26}$$

with the corresponding MSE defined as

$$\delta_B = E(\widehat{y}_{Bt}^{ns} - y_t^{ns})^2. \tag{4.27}$$

Next, the second regime A is specified; namely,

$$\widehat{y}_{At}^{ns} = E[y_t^{ns} | \{y_t\}_{t=-\infty}^{+\infty}] = E[y_t^{ns} | \{y_t^{ns} + y_t^s\}_{t=-\infty}^{+\infty}], \tag{4.28}$$

with the corresponding MSE defined as

$$\delta_A = E(\widehat{y}_{At}^{ns} - y_t^{ns})^2. \tag{4.29}$$

If we rely on our "nominal" and "real" example, extraction regime B would correspond to the common practice of deflating series after seasonal adjustment. Conversely, with extraction regime A, deflation occurs prior to seasonal adjustment.

Assumption A.10: The two extraction regimes specified above are obtained by means of linear filtering; that is,

$$\widehat{y}_{Bt}^{ns} \equiv v_1(L)x_t - v_2(L)z_t, \tag{4.30}$$
$$\widehat{y}_{At}^{ns} \equiv v_3(L)y_t, \tag{4.31}$$

where $v_j(L)$ are characterized as $v_j(L) \equiv \sum_{i=-\infty}^{+\infty} v_{ji}L^i$ with $\sum_{i=-\infty}^{+\infty} v_{ji}^2 < +\infty$, $j = 1, 2$, and 3. Finally, we can express the MSE of the two extraction regimes in terms of their linear filters:

$$\delta_A \equiv \delta_A(v_3), \tag{4.32}$$
$$\delta_B \equiv \delta_B(v_1, v_2). \tag{4.33}$$

Now we get to the basic question more formally stated: Given the design of the three filters $v_j(L)$, $j = 1, 2$, and 3, should we prefer $\widehat{y}_{Bt}^{ns}$ or $\widehat{y}_{At}^{ns}$? This question is answered by providing necessary and sufficient conditions guaranteeing that either δ_A or δ_B is smaller. Theorem 4.1 reproduced from Ghysels (1997a) conveniently summarizes the conditions for any given set of filters $v_j(L)$ and processes $\{x_t\}$, $\{y_t\}$, and $\{z_t\}$ satisfying the regularity assumptions stated before. This means, for instance, that the pairs $(\Delta y_t^{ns}, \Delta z_t^{ns})$ and $(\Delta_{(s)} y_t^s, \Delta_{(s)} z_t^s)$ may be correlated in any general way. Besides the general case, we shall consider several special cases such as uniform linear filtering and optimal MSE filtering. The main result can be stated as follows.

Theorem 4.1: Let x_t, y_t, and z_t be a triplet of stochastic processes with $x_t = y_t + z_t$ satisfying Assumptions A.8 and A.9. Moreover, let $v_j(L)$, $j = 1, 2, 3$ be linear filters corresponding to the extraction regimes and satisfying

Assumption A.10. Then, $\delta_A(\nu_3) \leq \delta_B(\nu_1, \nu_2)$ if and only if

$$
\int_{-\pi}^{\pi} \left\{ \left| \frac{1 - \nu_1(e^{-i\omega})}{(1 - e^{-i\omega})^k} \right|^2 f_y^{ns}(e^{-i\omega}) + \left| \frac{\nu_1(e^{-i\omega})}{1 + e^{-i\omega} + \cdots + e^{-11i\omega}} \right|^2 f_y^s(e^{-i\omega}) \right.
$$

$$
+ \left| \frac{\nu_2(e^{-i\omega}) - \nu_1(e^{-i\omega})}{(1 - e^{-i\omega})^{\tilde{k}}} \right|^2 f_z^{ns}(e^{-i\omega}) + \left| \frac{\nu_2(e^{-i\omega}) - \nu_1(e^{-i\omega})}{1 + e^{-i\omega} + \cdots + e^{-11i\omega}} \right|^2
$$

$$
\times f_z^s(e^{-i\omega}) + 2 \operatorname{RE}\left[\left(\frac{1 - \nu_1(e^{-i\omega})}{(1 - e^{-i\omega})^k} \right) \left(\frac{\nu_2(e^{+i\omega}) - \nu_1(e^{+i\omega})}{(1 - e^{+i\omega})^{\tilde{k}}} \right) \right]
$$

$$
\times f_{yz}^{ns}(e^{-i\omega}) - 2 \operatorname{RE}\left(\frac{\nu_1(e^{-i\omega})}{1 + e^{-i\omega} + \cdots + e^{-11i\omega}} \right)
$$

$$
\times \left(\frac{\nu_2(e^{+i\omega}) - \nu_1(e^{+i\omega})}{1 + e^{+i\omega} + \cdots + e^{+11i\omega}} \right) f_{yz}^s(e^{-i\omega}) - \left[\left| \frac{1 - \nu_3(e^{-i\omega})}{(1 - e^{-i\omega})^k} \right|^2 f_y^{ns}(e^{-i\omega}) \right.
$$

$$
\left. + \left| \frac{\nu_3(e^{-i\omega})}{1 + e^{-i\omega} + \cdots + e^{-11i\omega}} \right|^2 f_y^s(e^{-i\omega}) \right] \right\} d\omega \geq 0. \tag{4.34}
$$

Optimal filters obviously outperform ad hoc filters such as X-11 in terms of MSE when the objective is univariate seasonal adjustment filter design. Is this still true if seasonal adjustment is not the only data transformation? The answer is still affirmative, but it will crucially depend on having seasonal adjustment and the other data transformations done *in the right sequence*. Hence, we can face situations in which $\delta_B(\nu_1, \nu_2) \leq \delta_A(\nu_3)$ and situations in which $\delta_A(\nu_3) \leq \delta_B(\nu_1, \nu_2)$. How does the MSE of the uniform filter, δ_U, relate to this? One thing we know for sure is that putting seasonal adjustment last in the sequence and applying an optimal filter will always be better than uniform filtering – namely, $\delta_A(\nu_3) \leq \delta_U$. A statistical agency would, of course, then have to second guess the desired transformations before performing optimal signal-extraction filtering to produce the right $\delta_A(\nu_3)$. Because $\delta_A(\nu_3) \leq \delta_U$, one faces three possible outcomes when comparing optimal and uniform signal-extraction filters; namely,

$$
\delta_A(\nu_3) \leq \delta_U \leq \delta_B(\nu_1, \nu_2), \tag{4.35}
$$

$$
\delta_A(\nu_3) \leq \delta_B(\nu_1, \nu_2) \leq \delta_U, \tag{4.36}
$$

$$
\delta_B(\nu_1, \nu_2) \leq \delta_A(\nu_3) \leq \delta_U. \tag{4.37}
$$

The last case is directly related to the condition of Theorem 4.1 by using optimal extraction filters. Indeed, when the nonnegativity of (4.34) with optimal filtering is violated, we know that δ_B is best among all three alternatives. Typically, this does not happen very often, in particular when one starts to focus on cases in which $z_t^s \equiv 0$. The two most interesting cases are (4.35) and (4.36). The best strategy is $\delta_A(\nu_3)$ in both cases, but now we face the dilemma of deciding whether uniform filtering is second best – that is, better than optimal filtering applied to the most inefficient sequence of data transformations. Please note that this most inefficient sequence of data transformations is the currently institutionalized one except that optimal signal-extraction filters would be used instead of a program such as X-11. We can again exploit the arguments of Theorem 4.1 to settle the conditions for comparing (4.35) and (4.36).

Corollary 4.1: Let the assumptions of Theorem 4.1 hold and let $\delta_A(\nu_3) \leq \delta_B(\nu_1, \nu_2)$. Then uniform filtering yields a smaller MSE than $\delta_B(\nu_1, \nu_2)$; that is, $\delta_U \leq \delta_B(\nu_1, \nu_2)$ if and only if (4.34) holds with $\nu_1 \equiv \nu_1$, $\nu_2 \equiv \nu_2$, and $\nu_3 \equiv \nu_U$.

Naturally, Corollary 4.1 also applies when $z_t^s \equiv 0$ and hence $\nu_1(L) \equiv 1$. As a practical matter, it is very important to investigate how often one faces situations in which $\delta_U \leq \delta_B(\nu_1, \nu_2)$. To shed light on this, Ghysels (1997a) conducted several numerical calculations showing that most often δ_A dominates δ_B, and δ_U dominates δ_B as well.

To conclude this section, we would like to draw some comparisons with work by Geweke (1978), who studies temporal and sectoral aggregation of series combined with seasonal adjustment. The sectoral aggregation case studied by Geweke is most directly related to the theoretical discussion presented in this section. It concerns the problem of seasonally adjusting aggregate output (or any other series) from several industries when raw unadjusted series per industry are available. Hence, Geweke wondered whether a statistical agency should seasonally adjust all individual industries first and then aggregate the outcome or vice versa. This is quite similar to the extraction regimes A and B discussed. Geweke, in fact, considered a third regime, consisting of a multivariate extraction setup using z_t and x_t, *jointly* yielding an MSE that we denote δ_M. He showed that $\delta_M \leq \delta_A$ and δ_B, but the relationship between δ_A and δ_B was left unspecified. In this regard, the results presented here complement Geweke's analysis.

5 Estimation and Hypothesis Testing with Unfiltered and Filtered Data

5.1 Introduction

In his Foreword to this book, Tom Sargent addresses a fundamental issue pertaining to the econometric treatment of seasonality. One view is that all models are inherently misspecified, since they represent abstractions of a complex reality. When a dynamic optimization-based representative agent structural model is considered, one interpretation could be that the model is exclusively designed to capture business cycle fluctuations. In that context, it could be argued that to fit the model to actual data one should consider seasonally adjusted data to estimate the parameters of the model. This view is elegantly expressed by Sims (1974, 1993). Another view is to note that optimization-based structural models should reflect the decision-making process of representative agents facing seasonal fluctuations, and that such fluctuations are in fact an important information source regarding the dynamic propagation mechanism. In light of this view, it would be natural to estimate models with unadjusted data.

Although we cannot settle these issues in this chapter, we will consider some econometric consequences of using seasonally unadjusted and seasonally adjusted data. In a first section we examine the consequences of misspecifying the stochastic structure of seasonality in linear regression models. Hence, the first section deals with some of the pitfalls of using unadjusted data and getting the specification of the seasonality wrong. The remainder of the chapter deals with the consequences of using filtered data. We begin with the classical linear regression model, as examined by Sims (1974) and Wallis (1974). Then we expand on linear dynamic models, particularly ARMA and ARIMA models. Here our analysis is at first asymptotic, followed by an examination of finite sample results. We extend the univariate results to the case of cointegration. The chapter concludes with the intriguing debate about bias trade-offs and approximation errors alluded to by Sargent in the Forward to this book.

5.2   Linear Regression with Misspecified Seasonal Nonstationarity

There are two procedures that are commonly used to deal with seasonality in regression models involving unadjusted data. One consists of using seasonal dummies; the other relies on seasonal differences. The discussion in Chapters 2 and 3 revealed there might be some problems with either procedure, depending on whether unit roots at the seasonal frequency are present. In this first section, we examine the consequences of misspecifying the stochastic structure of seasonality in linear regression models. There are two types of misspecification that are the most common and that we will examine in this section. The first type of mispecification occurs when seasonal differences are taken when the processes in a regression are stationary but feature seasonal mean shifts. The second is the converse; that is, seasonal dummies are used when seasonal differencing ought to be applied.

We start with inappropriate seasonal differencing and consider a regression model similar to (1.1):

$$y_t = \sum_{s=1}^{S} \delta_{st} \gamma_s + z_t, \tag{5.1}$$

where $\phi(L)z_t = \theta(L)\varepsilon_t$ and $\varepsilon_t \sim$ i.i.d. $(0, \sigma^2)$ while $\theta(L)$ is a qth order polynomial and $\phi(L)$ is a pth order polynomial both with roots outside the unit circle. If the seasonality is deterministic, this is indeed the right regression to run. The corresponding seasonal difference model is

$$\Delta_S y_t = \Delta_S z_t. \tag{5.2}$$

The consequences of inappropriate seasonal differencing can be deduced from comparing (5.1) and (5.2). First, the seasonal differencing operator removes the seasonal means so that they are no longer identified. Second, the Δ_S filter results in $\Delta_S z_t$, which is a type of overdifferencing data; see Maravall (1995) for further details. If the true DGP is (5.1) with Δy_t replacing y_t, and hence the process is $I(1)$ with deterministic seasonality, then (5.2) becomes $\Delta_S y_t = \mu + (1 + L + \cdots + L^{S-1})z_t$. In this case also, the seasonal means are no longer identified and the consequences for the disturbances are similar to those of overdifferencing.

The second type of misspecification occurs when seasonal dummies are inappropriately used because the polynomial $\phi(L)$ has roots on the unit circle at seasonal frequencies. Abeysinghe (1991, 1994) and Franses, Hylleberg, and Lee (1995) pursue this by showing that estimation of the deterministic seasonality model of (5.1) can result in a high R^2 if the true DGP is a nonstationary seasonal process. Therefore, a high R^2 is not evidence in favor of deterministic seasonality and, similarly, conventional statistical evidence that the coefficients γ_s in (5.1) are nonzero can be obtained even if the process contains no deterministic component at all. This phenomenon is termed spurious deterministic seasonality.

To investigate this phenomenon, assume that the true DGP is the seasonal random walk of (2.19) with starting values $y_{s0} = 0$ for $s = 1, \ldots, S$. Thus, $E(y_{s\tau}) = 0$ for $s = 1, \ldots, S$, $\tau = 1, \ldots, T_\tau$. Ordinary least-squares estimation of the deterministic seasonal model (5.1) yields

$$\widehat{\gamma}_s = \frac{1}{T_\tau} \sum_{\tau=1}^{T_\tau} y_{s\tau}, \tag{5.3}$$

so that the dummy variable coefficient value is estimated by the sample mean of the observations for the relevant season. However, under the seasonal random walk DGP, $\widehat{\gamma}_s$ does not converge asymptotically to any fixed number. Indeed, even though the seasonal random walk DGP implies that $\gamma_s = 0$, $\widehat{\gamma}_s$ diverges, since (3.28) implies that $\sum_{\tau=1}^{T_\tau} y_{s\tau}$ must be divided by $T_\tau^{3/2}$ in order to obtain a well-defined asymptotic distribution. Thus, as the sample size increases, the estimated seasonal dummy variable coefficient will tend to get further away from its true value of zero.

Further, the R^2 value from the regression will be

$$R^2 = \left[\sum_{t=1}^{T} (\widehat{y}_t - \overline{y})^2 \right] \Big/ \left[\sum_{t=1}^{T} (y_t - \overline{y})^2 \right]$$

$$= \left[\sum_{t=1}^{T} \widehat{y}_t^2 - T\overline{y}^2 \right] \Big/ \left[\sum_{t=1}^{T} y_t^2 - T\overline{y}^2 \right],$$

where the fitted value is $\widehat{y}_t = \widehat{y}_{s\tau} = \widehat{\gamma}_s$ and the overall sample mean is given by $\overline{y} = (1/S) \sum_{s=1}^{S} \widehat{\gamma}_s$. Thus, in this context

$$R^2 = \frac{\frac{1}{T_\tau} \sum_{s=1}^{S} \left(\sum_{\tau=1}^{T_\tau} y_{s\tau} \right)^2 - \frac{1}{ST_\tau} \left(\sum_{s=1}^{S} \sum_{\tau=1}^{T_\tau} y_{s\tau} \right)^2}{\sum_{s=1}^{S} \sum_{\tau=1}^{T_\tau} y_{s\tau}^2 - \frac{1}{ST_\tau} \left(\sum_{s=1}^{S} \sum_{\tau=1}^{T_\tau} y_{s\tau} \right)^2}. \tag{5.4}$$

After dividing the numerator and denominator of (5.4) by T_τ^2, the distributional results of Subsection 3.2.5 can be shown to imply that

$$R^2 \Rightarrow \frac{\sum_{s=1}^{S} \left[\int_0^1 W_s(r)\, dr \right]^2 - \frac{1}{S} \left[\sum_{s=1}^{S} \int_0^1 W_s(r)\, dr \right]^2}{\sum_{s=1}^{S} \int_0^1 [W_s(r)]^2\, dr - \frac{1}{S} \left[\sum_{s=1}^{S} \int_0^1 W_s(r)\, dr \right]^2}. \tag{5.5}$$

The interpretation of (5.5) may not be immediately transparent, but it does have a number of important implications. First, R^2 does not here asymptotically converge to a fixed number, but rather converges to a distribution that involves functions of the independent Brownian motions relating to $y_{s\tau}$ ($s = 1, \ldots, S$).

Second, because of this fact, a conventional F test for the null hypothesis $\gamma_1 = \cdots = \gamma_S$ is not valid. Rejection of the null hypothesis would be spurious, and hence the apparent deterministic seasonality would also be spurious.

These results, pointed out by Abeysinghe and by Franses et al. (1995), directly echo the spurious regression phenomenon noted first by Granger and Newbold (1974) and formally analyzed by Phillips (1986). Clearly, if the presence of deterministic seasonality cannot be examined by means of a conventional F statistic, then another methodology is required that remains valid when the underlying process could be a nonstationary stochastic seasonal one. Franses et al. (1995) conduct a Monte Carlo study that confirms their theoretical findings. Abeysinghe (1994) and da Silva Lopes and Montanes (1999) also provide further simulation as well as empirical support.

To conclude this section, note that while the regression appearing in (5.1) is not generic, it is easy to extend the results in this section to general linear regression models.

5.3   The Classical Linear Regression Model and Filtering

The generic regression we will consider in this section is dynamic and may also contain other regressors; hence we will examine the following regression model:

$$
y_t = \sum_{j=1}^{p} \alpha_j y_{t-j} + x_t'\beta + \delta_t + \varepsilon_t,
$$
$$
t = -\ell, \ldots, 0, 1, \ldots, T + k. \tag{5.6}
$$

Note that the sample starts at $t = -\ell$ and ends at $T + k$ to accommodate two-sided filtering of data. This generic setup, which may include a constant and trend, defined by δ_t, will be used in most of the remainder of this chapter with different assumptions regarding the regressors. In the simplest case no lagged dependent variables are present, while the regressors in x_t are assumed to be nonstochastic.[1] This classical linear regression model will yield exact small sample results regarding the effect of filtering when lagged dependent variables are omitted and the disturbances are Gaussian. Next we will move to cases that are essentially AR(p) models, or more generally ARMA (p, q). Here the large sample behavior of the effects of filtering will be examined. Finally, we conclude with finite sample results for one specific case of a dynamic regression, in which it is assumed that the regressors are nonstochastic and that the error process $\varepsilon' = (\varepsilon_{-\ell}, \ldots, \varepsilon_{T+k})$ has a known covariance matrix, that is, $E(\varepsilon\varepsilon') = \Omega$. The covariance matrix does not have to be diagonal; hence we can allow for an MA error process, for instance.

---

[1] The vector of regressors x_t is assumed to include a constant.

We will assume that data are filtered with a *known and linear* filter. The most prominent example is the linear approximation to the X-11 filter discussed in the previous chapter. Since we restrict ourselves to linear filters, we exclude from our analysis the effect of the potential sources of nonlinearity in some of the filtering procedures such as X-11 and X-12 that were described in the previous chapter. The filter weights of the possibly two-sided filter are represented by the vector $(v_{-\ell}, \ldots, v_k)$. On the basis of this vector, we define the $(T + 1) \times (T + k + \ell + 1)$ filter matrix F as

$$
F = \begin{pmatrix}
v_{-\ell} & \cdots & \cdots v_k & & O \\
 & \ddots & & \ddots & \\
 & & \ddots & & \ddots \\
O & & v_{-\ell} & & v_k
\end{pmatrix}.
\tag{5.7}
$$

This matrix transforms a sample of size $\mathcal{T} = T + k + \ell + 1$ into a filtered data set with $T + 1$ observations, since at each end of the sample data are discarded by the two-sided filter.

Assumption 5.1: The filter weights $(v_{-\ell}, \ldots, v_k)$ are such that the following type of regressors are invariant to filtering $(X_F = X)$: (1) constant, (2) time trend. Moreover, the filter F applied to seasonal dummies yields a matrix of constants equal to the sum of seasonal mean shifts.

A filter that is symmetric and two sided, that is, with filter weights satisfying $v_{-i} = v_i$, and with weights that sum to one, that is, $\Sigma_{i=-m}^{m} v_i = 1$ where $m = k = \ell$, satisfies Assumption 5.1. Sometimes we will represent the filter as a polynomial lag operator, that is, $v(L) = \Sigma_{i=-\ell}^{k} v_i L^i$. Then a filter featuring $v(L) = v(L^{-1})$ and $v(1) = 1$ satisfies Assumption 5.1. On the basis of (5.6) and (5.7), we can define filtered and unfiltered data sets as $y' = (y_{-\ell}, \ldots, y_{T+k})$, $\mathcal{T} \times 1$, $y_F = Fy$, $(T + 1) \times 1$ and $y_U = Uy$, $(T + 1) \times 1$, where the matrix $U \equiv (O_{(T+1)\times\ell} I_{T+1} O_{(T+1)\times k})$ cuts away observations at each end of the sample, so that y_F and y_U are filtered and unfiltered data sets of equal sizes $T + 1$ with disturbances drawn from the same random vector ε.[2] Each element of y_F and

---

[2] While this setup is relatively simple, it should be noted that it is not unrealistic. In particular, one can think of a situation in which a statistical agency releases data simultaneously, unadjusted and adjusted, covering the same sample period, while the agency has actually more raw data available than is released to the public. The setup also has another advantage that must be emphasized at this stage. The filter matrix F appearing in (5.7) is not a square matrix. Consequently, filtering will always entail information loss that cannot be recovered or undone by an estimation procedure such as generalized least squares, which would essentially amount to taking the inverse of the matrix F.

y_U will be denoted y_t^F and y_t^U, respectively, $t = 0, \ldots, T$. Finally, we can write $\varepsilon_F = F\varepsilon$, $\varepsilon_U = U\varepsilon$, $X_U = UX$, and $X_F = FX$, where $X' = (x'_{-\ell}, \ldots, x'_{T+k})$.

The fact that Ω and F are known matrices implies that the covariance matrices of the filtered and unfiltered errors are also known, namely

$$E(\varepsilon_U \varepsilon'_U) \equiv \Omega_U = U\Omega U', \tag{5.8}$$

$$E(\varepsilon_F \varepsilon'_F) \equiv \Omega_F = F\Omega F'. \tag{5.9}$$

Sims (1974) and Wallis (1974) studied regression models without lagged dependent variables; hence they focused on a special case of (5.6), namely

$$y_t = x'_t \beta + \varepsilon_t, \qquad t = -\ell, \ldots, 0, 1, \ldots, T + k. \tag{5.10}$$

Using classical linear regression results, we have

$$\widehat{\beta}_j = (X'_j X_j)^{-1} X'_j y_j,$$

where $j = U$ or F and $\widehat{\beta}_j$ is the OLS estimator.

When regressors are nonstochastic and the disturbances in (5.10) are Gaussian, we have

$$E(\widehat{\beta}_F) = (X'_F X_F)^{-1} X'_F E(X_F \beta + F\varepsilon) = \beta,$$

and therefore the estimator is unbiased, whereas

$$\text{Var}(\widehat{\beta}_F) = (X'_F X_F)^{-1} X'_F \Omega_F X_F (X'_F X_F)^{-1},$$

which characterizes the finite sample distribution of the OLS estimator. Hence, we recover the standard results regarding the use of OLS in regressions with autocorrelated residuals. We can obviously extend these results outside the realm of fixed regressors and homoskedastic prefiltered errors. Instead of finite sample F tests, we would have asymptotic χ^2 tests and $\widehat{\beta}_F$ would be a consistent estimator. It is in fact the context in which Sims (1974) and Wallis (1974) portrayed their analysis.[3] A straightforward application of generalized least squares (GLS) would remedy the problem of computing standard errors and calculating test statistics. Hylleberg (1986) provides a more elaborate discussion of the classical Sims and Wallis result.

5.4 Filtering of ARIMA Models

In this section, we analyze the large sample behavior of OLS estimators in dynamic regression models, in particular ARIMA types of models. The analysis follows closely Ghysels and Perron (1993). We will put most of our emphasis on

[3] To be more accurate we should note that Sims (1974) considered a regression model that is slightly more general than (5.6), though without lagged dependent variables or deterministic trends. In Section 7 of this chapter we will consider the Sims (1974) setup.

the asymptotic bias, that is, the difference between the expected value of the limiting random variable with and without filtering. Since we are dealing with time series regressions involving possible nonstationary processes, we will start by considering a slight variation on the regression model appearing in (5.1), namely

$$y_t = \mu + \gamma t + z_t, \tag{5.11}$$

where we have again that $\phi(L)z_t = \theta(L)\varepsilon_t$ and $\varepsilon_t \sim$ i.i.d. $(0, \sigma^2)$. Here $\theta(L)$ is a qth order polynomial without roots on the unit circle and $\phi(L)$ is a pth order polynomial in L with at most one root on the unit circle and all other roots outside the unit circle. Hence, we are allowing for the possibility that $\{y_t\}$ represents a difference stationary process. In that case, (5.11) can be written as

$$y_t = \gamma + y_{t-1} + z_t, \tag{5.12}$$

where $z_t = \phi^*(L)^{-1}\theta(L)\varepsilon_t$ and $\phi^*(L)$ is defined such that $(1 - L)\phi^*(L) = \phi(L).$[4]

Like in the previous section, we denote by $v(L)$ a general linear two-sided filter.[5] As noted before, given the structure of the filters described in Assumption 5.1, we consider only polynomials $v(L)$ that are symmetric and whose coefficients sum to one, that is, for which $v(L) = v(L^{-1})$ and $v(1) = 1$. A symmetric filter with $v(1) = 1$ leaves unchanged the deterministic nonseasonal part of the series. Therefore, without loss of generality, we specify $\mu = \gamma = y_0 = 0$ in (5.11) and (5.12). Under these conditions, the filtered version of y_t, denoted $y_t^F = v(L)y_t$, is given by:

$$y_t = z_t, \qquad y_0 = 0, \tag{5.13}$$

$$y_t^F = v(L)z_t, \qquad y_0^F = 0. \tag{5.14}$$

We first consider regressions without estimated intercept and trend and then more general cases involving estimated trend and/or intercept. Hence we initially consider

$$y_t = \alpha y_{t-1} + \varepsilon_t, \tag{5.15}$$

$$y_t = \alpha y_{t-1} + \sum_{j=1}^{k} c_j \Delta y_{t-j} + \varepsilon_t, \tag{5.16}$$

where the estimator of the autoregressive parameter is specifically defined in the case of (5.15) as

$$\widehat{\alpha} = \sum_1^T y_t\, y_{t-1} \bigg/ \sum_1^T y_{t-1}^2 \tag{5.17}$$

---

[4] Note that, given our assumptions, $\phi^*(L)$ has all its roots outside the unit circle. It should also be noted that z_t is not the same stochastic process in (5.11) and (5.12).

[5] We have again the X-11 filter particularly in mind; i.e., $v(L)$ corresponds to (4.9). The setup is at first generic, however, like the matrix F, and also applies to linear filters other than X-11.

with unfiltered data, and with filtered data as

$$\widehat{\alpha}_F = \sum_1^T y_t^F y_{t-1}^F \Big/ \sum_1^T \left(y_{t-1}^F\right)^2. \tag{5.18}$$

In a first subsection we will examine the limiting behavior of $\widehat{\alpha}$ and $\widehat{\alpha}_F$ when $\phi(L)$ has a unit root; then in a second subsection we move on to stationary processes.

5.4.1 Filtering of Unit Root Processes

Both estimators converge to 1 as the sample size increases when $\phi(L)$ has a root on the unit circle. Hence, there is no asymptotic bias when the autoregressive model contains a unit root. To see this, note that from (5.12) with $\gamma = 0$, we have

$$y_t^F = v(L)y_t = v(L)y_{t-1} + v(L)z_t,$$
$$= y_{t-1}^F + w_t, \tag{5.19}$$

where $w_t = v(L)\phi^*(L)^{-1}\theta(L)\varepsilon_t$, which is a stationary and invertible ARMA process since $\phi^*(L)$ and $\theta(L)$ have roots outside the unit circle. Therefore, $T^{-1}\Sigma_1^T y_{t-1}^F w_t \to 0$ (in probability) as $T \to \infty$ and from (5.18), $(\widehat{\alpha}_F - 1) = T^{-1}\Sigma_1^T y_{t-1}^F w_t / T^{-1}\Sigma_1^T (y_{t-1}^F)^2$, which yields $\widehat{\alpha}_F \to 1$ (in probability). Finally, $\widehat{\alpha}$ also converges to 1 by using usual arguments [e.g., Phillips (1986)].

It should be noted, however, that the asymptotic distributions of $T(\widehat{\alpha}_F - 1)$ and $T(\widehat{\alpha} - 1)$ are different because of the transformed correlation structure of the error term. The variances of the processes $\{z_t\}$ and $\{w_t\}$ are, however, different. To see this note that $S_w^2 \equiv \text{Var}(w_t) = \text{Var}(v(L)z_t) = \Sigma_{j=-m}^m \Sigma_{l=-m}^m v_j v_l$ $\text{Cov}_z(j - 1)$ [see, e.g., (5.20) below], where $\text{Cov}_z(k)$ is the autocovariance function of $\{z_t\}$ at lag k. Hence, the limiting distribution of $T(\widehat{\alpha} - 1)$ is $(1/2)(W(1)^2 - S_z^2/\sigma_z^2)/\int_0^1 W(r)^2 \, dr$ and the limiting distribution of $T(\widehat{\alpha}_F - 1)$ is $(1/2)$ $(W(1)^2 - S_w^2/\Sigma_w^2)/\int_0^1 W(r)^2 \, dr$. The expressions for the limiting distributions change accordingly when an intercept and/or trend are estimated.

While the asymptotic distribution of $T(\widehat{\alpha} - 1)$ and $T(\widehat{\alpha}_F - 1)$ are different, the asymptotic distribution of the Dickey–Fuller and Phillips–Perron statistics are identical [and as tabulated in Fuller (1996)]. This is because the structure of the errors in each case is such that the corrections applied by each procedure effectively eliminate the dependency of the asymptotic distribution on nuisance parameters associated with the correlation in the errors. Both z_t and w_t satisfy the mixing conditions for the application of a functional central limit theorem to their partial sums [see, e.g., Phillips (1986) and Phillips and Perron (1988)]. Let $\sigma_z^2 = \lim_{T\to\infty} T^{-1} E[S_{T,z}^2]$, where $S_{T,z} = \Sigma_1^T z_t$; similarly, $\sigma_w^2 = \lim_{T\to\infty} T^{-1} E[S_{T,w}^2]$, where $S_{T,w} = \Sigma_1^T w_t$. We note that $\sigma_z^2 = 2\pi h_z(0)$ and $\sigma_w^2 = 2\pi h_w(0)$, where $h_x(0)$ denotes the nonnormalized spectral density function of x evaluated at frequency 0. We have $\sigma_w^2 \equiv \sigma_\varepsilon^2[v(1)^2\theta(1)^2/\phi^*(1)^2] =$

$\sigma_\varepsilon^2[\theta(1)^2/\phi^*(1)^2] \equiv \sigma_z^2$, using the fact that $v(1) = 1$ since the weights of the filter sum to unity. Clearly, the property $v(1) = 1$ plays a key role here. It will also be crucial when we discuss cointegration and filtering later in this chapter.

The basic result that the asymptotic null distributions of the unit root test statistics are not affected by filtering the data does not imply, however, that the finite sample distributions are unaffected. The simulation results reported in Ghysels and Perron (1993) show that, indeed in some cases, the finite sample distributions can be substantially different after filtering.

It is worth noting that the unit root autoregressive model yields results that are *stronger* than those obtained for the classical linear regression model. Indeed, we showed in the previous section that while $\widehat{\beta}_F$ remained unbiased, its co-variance matrix was affected by filtering. Consequently, filtering has an impact on the (asymptotic) distribution of test statistics. In the case of unit root tests, this does not apply because the covariance structure of the stationary part vanishes asymptotically.

5.4.2 Filtering Stationary ARMA Processes

When the polynomial $\phi(L)$ does not contain a unit root, things are rather different because there exists an asymptotic bias induced by filtering, and – more importantly – Ghysels and Perron (1993) show that the bias is positive in the case of the linear X-11 filter. Thus, in (5.17) and (5.18), $\widehat{\alpha}_F$ has a limit greater than that of $\widehat{\alpha}$. Hence, because of the upward bias one can expect unit root tests performed with filtered data to be less powerful against stationary alternatives.

To clarify this, let us consider the limiting behavior of $\widehat{\alpha}$ in (5.17); in particular, we have $p \lim_{T\to\infty} \widehat{\alpha} = p \lim_{T\to\infty} T^{-1}\Sigma_1^T y_t\, y_{t-1}/p \lim_{T\to\infty} T^{-1}\Sigma_1^T y_{t-1}^2 = \gamma_y(1)/\gamma_y(0)$, where $\gamma_y(j)$ is the autocovariance function of $\{y_t\}$ at lag j. With filtered data, we have $p \lim_{T\to\infty} \widehat{\alpha}_F = p \lim_{T\to\infty} T^{-1}\Sigma_1^T y_t^F y_{t-1}^F/p \lim_{T\to\infty} T^{-1}(\Sigma_1^T y_{t-1}^F)^2$. For the more general case that will follow, consider the probability limit of $T^{-1}\Sigma_1^T y_t^F y_{t-s}^F$. Denoting the order of the symmetric MA polynomial $v(L)$ by m, we have

$$p \lim_{T\to\infty} T^{-1} \sum_1^T y_t^F y_{t-s}^F = p \lim_{T\to\infty} T^{-1} \sum_1^T [v(L)y_t][v(L)y_{t-s}],$$

$$= p \lim_{T\to\infty} T^{-1} \sum_1^T \sum_{i=-m}^m v_i\, y_{t+i} \sum_{j=-m}^m v_j\, y_{t+j-s},$$

$$= \sum_{i=-m}^m \sum_{j=-m}^m v_i v_j \left[p \lim_{T\to\infty} T^{-1} \sum_1^T y_{t+i}\, y_{t+j-s} \right],$$

$$= \sum_{i=-m}^m \sum_{j=-m}^m v_i v_j \gamma_y(i - j + s) \equiv \gamma_y^F(s).$$

$$(5.20)$$

Using (5.18)–(5.20), we find it easy to deduce the following asymptotic bias of $\widehat{\alpha}_F$, which we denote as $b(\widehat{\alpha}_F, 0)$ for reasons that will become clear shortly:

$$b(\widehat{\alpha}_F, 0) = [\gamma_y^F(1)/\gamma_y^F(0)] - [\gamma_y(1)/\gamma_y(0)]. \tag{5.21}$$

Using (5.21), we can numerically compute the asymptotic bias, provided we specify the covariance structure of the original series $\{y_t\}$, that is, the prefiltered series. Before we do so, we now consider the more general case of the AR(p) process (5.16) with $p = k + 1$, where now α represents the sum of the autoregressive coefficients.

First write (5.16) as

$$y_t = \phi_1 y_{t-1} + \phi_2 y_{t-2} + \cdots + \phi_p y_{t-p} + \varepsilon_t, \tag{5.22}$$

where, in accordance with (5.16), we have $\alpha = \Sigma_{j=1}^p \phi_j$ and $c_i = -\Sigma_{j=1}^{p-1}\phi_j$. Let $\phi' = [\phi_1, \phi_2, \ldots, \phi_p]$. The least-squares estimator of the vector of parameters ϕ is given by $\widehat{\phi} = A_n^{-1} V_n$, where $A_n = T^{-1}\Sigma_{t=p+1}^T X_t' X_t$, $V_n = T^{-1}\Sigma_{t=p+1}^T X_t' y_t$, where $X_t = (y_{t-1}, \ldots, y_{t-p})$ is a $1 \times p$ vector of lagged values of the data. For the filtered data, $\widehat{\phi}_F$ is defined similarly with $\{y_{t-s}\}_{s=0}^p$ replaced by $\{y_{t-s}^F\}_{s=0}^p$. Let $e' = (1, 1, \ldots, 1)$ be a $1 \times p$ vector of ones. Then $\widehat{\alpha}_F = e'\widehat{\phi}_F$ and $\widehat{\alpha} = e'\widehat{\phi}$. Denote the asymptotic bias of $\widehat{\alpha}_F$ in (5.21) when k lags of first differences of the data are included by $b(\widehat{\alpha}_F, k)$ (note that $p = k + 1$). Then for $\{y_t\}$ generated by (5.22), we have

$$b(\widehat{\alpha}_F, k) \equiv p\lim_{T\to\infty} \widehat{\alpha}_F - p\lim_{T\to\infty} \widehat{\alpha}$$
$$= e'A^{-1}V - e'A_F^{-1}V_F, \tag{5.23}$$

where A is a $p \times p$ matrix with elements $a_{ij} = \gamma_y(i - j)(i, j = 1, \ldots, p)$, V is a $p \times 1$ vector with elements $v_i = \gamma_y(i)(i = 1, \ldots, p)$; A_F and V_F are defined similarly with elements $a_{ij}^F = \gamma_y^F(i - j)$ and $v_i^F = \gamma_y^F(i)$, with $\gamma_y^F(s)$ defined by (5.20).

The expression in (5.23) provides a computable expression for the limiting bias of the least-squares estimator of $\widehat{\alpha}_F$ in (5.16) when k lags of first differences of the data are included (i.e., in a pth order autoregression with $p = k + 1$). Given (5.20), all that is needed to compute this asymptotic bias is the covariance function $\gamma_y(s)$ of the original unfiltered data and the weights of the filter polynomial $v(L)$. Note that when $k = 0$, $p = 1$ and (5.23) reduces to (5.21).

Though the asymptotic bias as described in (5.23) pertains to the estimators $\widehat{\alpha}_F$ when no constant and trend are included in the regression, the same quantitative result holds if a constant or a constant and a time trend are included in the regression [see Proposition 2 in Ghysels and Perron (1993)].

Ghysels and Perron quantify the nonvanishing asymptotic bias for a variety of parametric specifications when filtered data are used. The bias calculations appearing in Ghysels and Perron have some interesting practical implications for unit root tests, since they show that the biases in the sum of the autoregressive coefficients are substantial unless the order of the estimated autoregression is at least as great as the seasonal period when correlation at seasonal lags is present in the original data and the X-11 filter is applied. The practical implication for unit root tests of the Dickey–Fuller type is that one must consider an estimated autoregression of an order at least as great as the seasonal period in order to avoid a substantial bias. Hence, seasonal adjustment of the data does not permit a reduction in the order of the autoregression (also see the discussion of Subsection 3.3.4). This is basically because even though seasonal adjustment eliminates correlation in the data at seasonal frequencies (albeit imperfectly), it induces a bias in the autocorrelation function at lags less than the seasonal period, which does not vanish even asymptotically.

In most practical applications of Dickey–Fuller (1979, 1981) tests with seasonally adjusted data, one does not follow such a strategy but considers instead short lags because it is believed that seasonal adjustment removes seasonal correlation and does not produce any side effects. Let us note too that even with an order greater than the seasonal, there will still remain an upward bias. Given that an autoregression of an order at least as great as the seasonal period is needed with filtered data, it is reasonable to expect higher power of a test for a unit root with unfiltered data than with filtered data. Indeed, not only does the latter not permit a reduction in the length of the autoregression estimated (which would help increase power), but it induces an upward bias in the statistic of interest.

5.5   Finite Sample Approximations of Filtering Effects

The purpose of this section is to further examine the effect of filtering on autoregressive models. Ghysels and Perron (1993) used Monte Carlo simulation to study the small sample behavior. In the case of AR(1) models with possibly nonstochastic regressors, we can in fact study the small sample distributions and isolate the filtering effects through approximate analytical methods. Consider again the dynamic regression model:

$$y_t = \alpha y_{t-1} + x'_t \beta + \varepsilon_t, \qquad t = -\ell, \ldots 0, 1, \ldots, T+k. \qquad (5.24)$$

The vector x_t may contain a constant, trend, and possibly other regressors. Using the filtering matrices F and U together with the error covariance matrix

Ω introduced in Section 5.3, we can formulate the ARX(1) using

$$R_\alpha = \begin{pmatrix} b & 0 & 0 & \cdots & 0 & 0 \\ b\alpha & 1 & 0 & \cdots & 0 & 0 \\ b\alpha^2 & \alpha & 1 & \cdots & 0 & 0 \\ \vdots & & & & & \\ & & & & & \\ b\alpha^{T-1} & \alpha^{T-2} & \alpha^{T-3} & & \alpha & 1 \end{pmatrix},$$

$$b = \begin{cases} (1-\alpha^2)^{-1/2} & \text{if } \alpha \in (-1,1) \\ 0 & \text{otherwise} \end{cases}. \tag{5.25}$$

So that

$$\text{Var}(y) \equiv \Omega^y = R_\alpha \Omega R'_\alpha, \tag{5.26}$$

$$\text{Var}(y_F) \equiv \Omega^y_F = F R_\alpha \Omega R'_\alpha F', \qquad (T+1) \times (T+1), \tag{5.27}$$

$$\text{Var}(y_U) \equiv \Omega^y_U = U R_\alpha \Omega R'_\alpha U', \qquad (T+1) \times (T+1). \tag{5.28}$$

To facilitate the discussion, we will denote the estimator of α when unfiltered data are used at $\widehat{\alpha}_U$ while, as in the previous secion, $\widehat{\alpha}_F$ denotes the estimator with filtered data. Let $D_1 \equiv [I_T \, O_{T \times 1}]$, $D_2 \equiv [O_{T \times 1} \, I_T]$, and $M_j \equiv D_2 X_j$, $j = F$ or U. Then the OLS estimator of α and β may be written as, for $j = F, U$,

$$\hat{\alpha}_j \equiv y'_j S_j y_j / y'_j B_j y_j, \tag{5.29}$$

$$\hat{\beta}_j \equiv (M'_j M_j)^{-1} M'_j (D_2 y_j - \hat{\alpha}_j D_1 y_j), \tag{5.30}$$

where

$$S_j \equiv 1/2(D'_1 A_j D_2 + D'_2 A_j D_1),$$

$$B_j \equiv D'_1 A_j D_1,$$

$$A_j \equiv I - X_j (X'_j X_j)^{-1} X'_j.$$

The OLS estimator $\hat{\alpha}_F$ has the obvious disadvantage that, apart from its small sample bias, it may not be consistent in large samples even if we start out with a white-noise innovation structure for the DGP of the raw data process. In other words, even though Ω may be diagonal, it is clear that Ω_F will typically be nondiagonal. Consequently, it is not clear when we compare $\hat{\alpha}_U$ and $\hat{\alpha}_F$ whether we are studying the effect of filtering or the combination of lagged dependent variables in a regression with autocorrelated residuals. Since it is assumed that Ω_F and Ω_U are known, with Ω, U, and F known, it is relatively straightforward to extend these results to GLS estimators.

The OLS estimators $\hat{\alpha}_j$, $j = F$ or U, are ratios of quadratic forms. Ghysels and Lieberman (1996) suggest using approximations to the moments of this ratio by means of the Laplace method, which has the appeal of providing relatively simple analytical expressions, involving only matrix operations. Moreover, with the Laplace approximation it is relatively easy to isolate the effect of filtering on the moments of the estimators $\hat{\alpha}_j$.

To study the Laplace approximation to say $b(\widehat{\alpha}_F, 0)$ appearing in (5.21), which we denote $b_L(\widehat{\alpha}_F, 0)$, we can write $b_L(\widehat{\alpha}_F, 0) \equiv \text{bias}_L(\widehat{\alpha}_F) - \text{bias}_L(\widehat{\alpha}_U)$. In all cases the subscript refers to the fact that the Laplace approximation is used. Then Ghysels and Lieberman (1996) show that

$$\text{bias}_L(\hat{\alpha}_j) = \frac{\mu_j'(S_j - \alpha B_j)\mu_j + tr[\Omega_j^y(S_j - \alpha B_j)]}{\mu_j' B_j \mu_j + tr(\Omega_j^y B_j)}, \tag{5.31}$$

where $\mu_j = j R_\alpha^* X' \beta$, with R_α^* the same as R_α except that b is replaced in the first column by $(1 - \alpha)^{-1}$.

Moreover, one can also calculate the Laplace approximation of the mean-squared error of $\widehat{\alpha}_j$, namely:

$$\text{MSE}_L(\hat{\alpha}_j) = \frac{[\mu_j' S_j \mu_j + tr(\Omega_j^y S_j)]^2 + 2[tr(\Omega_j^y S_j)^2 + 2\mu_j S_j \Omega_j S_j \mu_j]}{[\mu_j' B_j \mu_j + tr(\Omega_j^y B_j)]^2}$$
$$+ \alpha\left[\alpha - 2\frac{\mu_j' S_j \mu_j + tr(\Omega_j^y S_j)}{\mu_j' B_j \mu_j + tr(\Omega_j^y B_j)}\right], \qquad j = F, U. \tag{5.32}$$

It may be worth emphasizing that so far no restrictions have been imposed on the filter matrix F or on the nature of the exogenous regressors. Imposing restrictions on either one or both will greatly simplify the expressions in (5.31) and (5.32). One set of restrictions, arising fairly naturally in the design of filters, are those appearing in Assumption 5.1. Combined with the fact that the exogenous regressors contain only at most a linear trend and a constant, this yields the following set of assumptions.

Assumption 5.2: Let the filter matrix F satisfy Assumption 5.1 and the regressors be at most a constant and a linear trend.

Under Assumption 5.2, $S_j = S$, $B_j = B$, and $\mu_j = \mu$ for $j = U$ or F. While the assumption is not necessary to study small sample behavior, it does make it easier to obtain even simpler formulae. Let us first consider the case where $\mu = 0$ followed by that where the mean vector is nonzero. The former corresponds, of course, to the AR(1) model without intercept, while the latter is one where a constant and/or a linear trend are included.

Ghysels and Lieberman (1996) show that when Assumption 5.2 holds with $\mu = 0$, then the filtering effect on the bias of the estimator $\hat{\alpha}_F$ can be expressed as

$$\text{bias}_L(\hat{\alpha}_F) - \text{bias}_L(\hat{\alpha}_U)$$

$$= \frac{[vec(S' \otimes B')]'[F \otimes U] \otimes (F \otimes U) - (U \otimes F)(U \otimes F)vec(\Omega^y \otimes \Omega^y)}{[vec(B' \otimes B')]'[(F \otimes U) \otimes (F \otimes U)]vec(\Omega^y \otimes \Omega^y)}.$$

$$(5.33)$$

Moreover, $\text{bias}_L(\hat{\alpha}_F)$ is independent of T. Hence, with $\mu = 0$, the Laplace approximation becomes independent of T, a consequence of the design of the matrices S, B, and Ω^y in this particular case. Formula (5.33) can be used to compute the asymptotic bias.[6]

The following result covers the case where μ is no longer equal to the zero vector, namely

$$\text{bias}_L(\hat{\alpha}_F) - \text{bias}_L(\hat{\alpha}_U)$$

$$= \frac{\mu'(S - \alpha B)\mu}{\mu' B \mu} \left[\frac{1 + (vecS_1')'(F \otimes F)vec\Omega^y}{1 + (vecB_1')'(F \otimes F)vec\Omega^y} \right.$$

$$\left. - \frac{1 + (vecS_1')'(U \otimes U)vec\Omega^y}{1 + (vecB_1)'(U \otimes U)vec\Omega^y} \right] \qquad (5.34)$$

where $S_1 = (S - \alpha B)$, $[\mu'(S - \alpha B)\mu]^{-1}$, and $B_1 = B(\mu' B \mu)^{-1}$.

Ghysels and Lieberman (1996) found that the Laplace approximation strikes a balance between (1) simplicity of calculations and (2) accuracy of the approximations. Indeed, the relatively simple formulae, particularly those pertaining to the first-order Laplace expansion, seem to describe the effect of filtering very well both for the bias and MSE, with the exception of values for α near -1, that is, part of the parameter space which is of lesser practical importance.

5.6   Filtering and Cointegration

The effect of filtering on tests for cointegration is, for obvious reasons, closely related to the effects of filtering on tests for a unit root. Hence, it will not be surprising to find many similarities between the analysis in this section and the discussion appearing in Section 5.4.

Let us consider again the single equation setup of Subsection 3.6.1 in Chapter 3. In particular, let us reexamine the linear consumption function

---

[6] In fact, this formula is relatively simple (one can use small values of T) in comparison to those used by Ghysels and Perron (1993), described in the previous section, and Hansen and Sargent (1993), described in the next section.

involving income, denoted y_t, and aggregate consumption, c_t. We consider only zero-frequency cointegration since y_t^F and c_t^F no longer contain unit roots at the seasonal frequencies, at least for the class of filters such as the linear approximation of X-11.[7] Hence, the analysis focuses on comparing [using the notation of (3.87)]

$$w_t^{(1)} = c_t^{(1)} - k_1 y_t^{(1)},$$ (5.35)

with the cointegrating relationship applied to filtered data:

$$w_t^F = c_t^F - k_F y_t^F,$$

where $c_t^F = v(L)c_t$ and $y_t^F = v(L)y_t$.[8]

The questions of interest here are how filtering affects cointegration on the one hand and how filtering alters the cointegrating parameter; that is, how do k_1 and k_F differ? Ericsson, Hendry, and Tran (1994) analyze these questions. Similar to the analysis in Ghysels and Perron (1993), they impose restrictions on the filter weights, the same as those appearing in Assumption 5.1, and show that a symmetric two-sided filter with weights summing to one leaves cointegration relationships unaffected. Hence, like in unit root tests, filters satisfying $v(L) = v(-L)$ and $v(1) = 1$ leave zero-frequency inference unaffected, at least asymptotically. They also show that this result not only applies to a single equation cointegration setup but easily generalizes to vector cointegration approaches. Therefore, we obtain results that are stronger than those of Sims (1974) and Wallis (1974). Indeed, similar to the analysis of Ghysels and Perron (1993) pertaining to unit root tests, Ericsson, Hendry, and Tran (1994) find that cointegration tests are invariant to filtering; that is, not only consistency holds as in Sims (1974) and Wallis (1974), but the asymptotic distribution of the tests is unaffected.

It is perhaps worth noting parenthetically that $y_t^{(1)}$ and y_t^F also are expected to be cointegrated with cointegration vector $(1, -1)$. This property holds, provided one imposes again the usual assumptions on the filter $v(L)$. Ghysels, Granger, and Siklos (1996) examine this feature in fact to appraise nonlinearity in the X-11 seasonal adjustment program (cf. the discussion in Subsection 4.3.2).

Obviously, the aforementioned results hold asymptotically. In practice, it is clear that filtering may seriously interfere with tests for cointegration. Ericsson et al. (1994) illustrate that this is the case by examining an empirical example involving a money demand equation for the U.K.

---

[7] As noted in Chapter 4 and in Bell (1992), the linear X-11 filter removes unit roots at the seasonal frequencies.

[8] In a sense $c_t^{(1)}$ and $y_t^{(1)}$ are also filtered data constructed to isolate the zero frequency. Whenever we refer to filtered data, we will always consider the $v(L)$ filter.

5.7   Bias Trade-Offs and Approximation Errors

The arguments presented so far clearly point in the direction of model speci-fication, taking seasonality explicitly into account and estimating models with unadjusted data. Yet, estimating econometric models with unadjusted data also generates a number of nontrivial issues. In Section 5.2 we discussed the con-sequences of misspecifying the seasonal as deterministic when in fact it is stochastic and nonstationary and vice versa. Here we continue to study the consequences of misspecifying the nonseasonal.

The regression model considered by Sims (1974) can be written as follows [adopting Sims' (1974) notation]:

$$y_t^{ns} = x_t^{ns} * b^{ns} + \varepsilon_t^{ns}, \tag{5.36}$$

$$y_t^s = x_t^s * b^s + \varepsilon_t^s, \tag{5.37}$$

where ε_t^i is orthogonal to x_t^i for $i = ns, s$ and all the stochastic processes $\{y_t^i, x_t^i, \varepsilon_t^i, i = ns, s\}$ are covariance stationary. The notation $x_t^i * b^i$ stands for

$$x_t^i * b^i \equiv \sum_{k=-\infty}^{+\infty} b^i(k) x^i(t-k), \qquad i = ns, s. \tag{5.38}$$

Hence, compared to (5.6), we now have a two-sided projection of the regressors x_t into y_t. Naturally, one defines the observed raw data x_t and y_t as

$$y_t = y_t^{ns} + y_t^s,$$

$$x_t = x_t^{ns} + x_t^s,$$

yielding the linear regression

$$y_t = x_t * b + \varepsilon_t. \tag{5.39}$$

Sims (1974, 1993) points out that, despite all the arguments that would clearly lead toward the use of unadjusted data for econometric inference, one might be led to use seasonally adjusted data. Sims (1974) observed that, with the use of unadjusted data, standard inference procedures will often result in relatively close fits at seasonal frequencies and underemphasize other frequencies. The difference in fit results from a high concentration of spectral power in a small-frequency band. Hence, the loss function will put a heavy penalty on misfitting the seasonal frequencies. To clarify this point, consider the OLS estimator of the polynomial b of (5.39). This estimator minimizes a weighted average of the squared differences between the Fourier transforms $\mathbf{b}$ and $\tilde{b}$ of $\mathbf{b}$ and b (where $\mathbf{b}$ is the OLS estimator of b), with weights given by the spectral density of x_t, denoted S_x. In other words, the OLS minimizes the following quantity:

$$\int_{-\pi}^{\pi} S_x(\omega) |\widetilde{\mathbf{b}}(\omega) - \widetilde{b}(\omega)|^2 \, d\omega. \tag{5.40}$$

Because the difference $|\widetilde{\mathbf{b}}(\omega) - \widetilde{b}(\omega)|^2$ is given especially strong weight in the OLS criterion function, one expects a *bias trade-off across frequencies with a high penalty for misfitting the seasonal frequency.* Sims (1974) made this observation in the context of a simple (nonstructural) distributed lag model.

The argument has been extended recently to a very different (and more challenging) context of rational expectations models. Such models impose some nontrivial cross-frequency restrictions that make it difficult to disentangle directly the influence of seasonal and nonseasonal shocks. Except for the self-evident reasons showing that the use of adjusted data possibly introduces severe biases and discards important information about the dynamics [like the adjustment costs example in Ghysels (1988, 1994a)], Sims (1993) argues that using seasonal adjustment filters could in some cases produce better estimates for certain parameters. The argument is based on the fact that the loss function (5.40) overemphasizes the fit at the seasonal frequencies. With seasonal adjustment, other frequencies get a better representation in the loss function. The argument is essentially one of approximation errors, that is, errors caused by misspecifying the economic model, at the seasonal and other frequencies. Since the large spectral peaks at the seasonal frequencies overemphasize approximation errors at those frequencies, we may be worse off in estimating structural parameters not directly related or unrelated to seasonal fluctuations.

Hansen and Sargent (1993) develop a general apparatus to evaluate approximation errors in maximum likelihood based estimation and use it to evaluate Sims' arguments. The key result is a frequency-domain expansion for the criterion that maximum likelihood estimates (MLEs) of a misspecified model implicitly minimize. The fundamental link between maximum likelihood estimation and the Kullback and Leibler (1951) information criterion is used in the context of stationary and ergodic Gaussian linear vector time series models. More specifically, consider the process $\{Y_t\}$ with population mean μ and spectral density $F(\omega)$, where $-\pi \leq \omega \leq \pi$. Hansen and Sargent consider cases in which the econometrician fits an approximating model parameterized by the vector δ and produces a mean $\mu(\delta)$ and spectral density $G(w, \delta)$, $-\pi \leq \omega \leq \pi$. Using results from Hannan (1973), Dunsmuir and Hannan (1976), and Deistler, Dunsmuir, and Hannan (1978), Hansen and Sargent (1993) establish that the MLE of δ for the misspecified model converges almost surely to the minimizer of

$$A(\delta) \equiv A_1(\delta) + A_2(\delta) + A_3(\delta), \tag{5.41}$$

where

$$A_1(\delta) \equiv \frac{1}{2\pi} \int_{-\pi}^{\pi} \log \det G(\omega, \delta)\, d\omega,$$

$$A_2(\delta) \equiv \frac{1}{2\pi} \int_{-\pi}^{\pi} \operatorname{tr}\left[G(\omega, \delta)^{-1} F(\omega)\right] d\omega,$$

$$A_3(\delta) \equiv [\mu - \mu(\delta)]' G(0, \delta)^{-1} [\mu - \mu(\delta)].$$

When the model is correctly specified, then both $A_2(\delta)$ and $A_3(\delta)$ disappear. The last term, $A_3(\delta)$, vanishes when the mean is correctly specified.

Sims (1993) focused primarily on $A_2(\delta)$, suggesting the replacement of $F(\omega)$ by a seasonally adjusted spectral density $F_{sa}(\omega)$ in order to tone down sources of misspecification in $G(\omega, \delta)$, where ω lies in the seasonal frequency bands. Sims (1993) stresses examples in which seasonality is an exogenous source of fluctuations, such as, for instance, rational expectations models with seasonal characteristics for the endowment process. Namely, these are models where endogenous processes, such as consumption in a permanent income model or physical capital acquisition in an investment model, may be seasonal because of the exogenous driving processes, but would not generate seasonal variation otherwise. Hansen and Sargent (1993) go further in their analysis of sources of misspecification and consider cases in which the data are generated by a periodic model and the econometrician incorrectly fits an aperiodic model estimated either with unadjusted data, with or without seasonal mean shifts in the fitted model, or with seasonally adjusted data. They also consider cases in which the data are generated by an aperiodic seasonal model with endogenous and exogenous sources of seasonality, and the econometrician estimates misspecified models, incorrectly modeling the exogenous or endogenous sources of seasonality. Several of their examples reinforce the arguments advanced by Sims. The framework, however, is also able to generate cases that would go strongly against it as well.

So far, the discussion has exclusively emphasized biases of parameter estimates, leaving out issues of efficiency and testing. Quite often seasonal frequencies are extremely useful in identifying and estimating parameters of interest that determine the shape of the transfer function at other frequencies. The adjustment cost model discussed in Ghysels (1988) may serve as an example. In such a model, seasonal frequencies are very informative and help to pin down the adjustment cost parameter. Further research in this area is certainly going to be very stimulating in the years to come.

6   Periodic Processes

6.1   Introduction

As discussed in Chapter 1, periodic processes are processes in which the coefficients change with the seasons of the year. A deterministic seasonal process, in which the intercept changes seasonally, can be viewed as a special case of a periodic process. However, allowing other parameters to vary seasonally opens up a rich class of models that have been found to be useful as descriptions of economic variables; applications include Birchenhall et al. (1989), Flores and Novales (1997), Franses and Romijn (1993), Herwartz (1997), and Osborn and Smith (1989).

These processes have not been very widely applied in economics to date. Certainly their applications are fewer than those of either deterministic seasonality or nonstationary stochastic seasonality, with little analysis of the implications of seasonal adjustment for periodic processes. This explains why we examine periodic processes after other topics have been dealt with in some detail. Nevertheless, a number of studies show that periodic processes can arise naturally from the application of economic theory to modeling decisions in an economic context, and their role should not be dismissed as unimportant. For example, Gersovitz and MacKinnon (1978) and Osborn (1988) argue that a process of this type arises when modeling the seasonal decisions of consumers, while Hansen and Sargent (1993) suggest that it could also arise from seasonal technology. Once it is admitted that the underlying economic driving forces such as preferences or technologies may vary seasonally, then subtle periodic seasonal effects may come into play even in contexts generally considered to be nonseasonal. One such example is the study of Ghysels (1991, 1994b), who explores the periodic nature of U.S. business cycle turning points.

This chapter considers extensions of linear time series modeling issues to the periodic case. Some other (nonlinear) periodic effects are considered in Chapter 7. We begin the current chapter by presenting some simple periodic processes, thereby extending the introductory discussion of Chapter 1. We then move to a fuller examination of the properties and methods of analysis for periodic processes. In particular, Sections 6.4 and 6.5 examine issues concerning integration and cointegration for periodic processes.

6.2 Some Simple Periodic Processes

Earlier chapters have already used the double subscript notation, whereby the season $(s = 1, \ldots, S)$ and year $(\tau = 1, \ldots, T_\tau)$ are indicated for the time series observations over the total sample $t = 1, \ldots, T$. This notation is natural for periodic processes and has been used in this context by Pagano (1978), Troutman (1979), Tiao and Grupe (1980), and others. In this notation, a periodic ARMA (p, q) process has the general form

$$\phi_s(L)y_{s\tau} = c_s + \theta_s(L)\varepsilon_{s\tau}, \qquad s = 1, \ldots, S, \ \tau = 1, \ldots, T_\tau, \quad (6.1)$$

where $\phi_s(L) = 1 - \phi_{s1}L - \cdots - \phi_{sp}L^p$ and $\theta_s(L) = 1 - \theta_{s1}L - \cdots - \theta_{sq}L^q$ are polynomials in the conventional lag operator L and $\varepsilon_{s\tau}$ is an i.i.d. process over both season and year. Thus, $E(\varepsilon_{s\tau}\varepsilon_{kj}) = 0$ unless $s = k$ and $\tau = j$. Heteroskedasticity over seasons is typically permitted, so that $E(\varepsilon_{s\tau}^2) = \sigma_s^2$. To avoid later confusion, it should be emphasized that L in (6.1) operates on the season, so that the one-period lagged observation is $Ly_{s\tau} = y_{s-1,\tau}$ for $s = 2, \ldots, S$; applied to the first observation of year τ, $Ly_{1\tau} = y_{S,\tau-1}$. Further, note that $y_{s\tau}$ for $s \leq 0$ is understood to refer to the appropriately lagged value $y_{S+s,\tau-1}$. The annual lag value of $y_{s\tau}$ is $L^S y_{s\tau} = y_{s,\tau-1}$.

The polynomial orders p and q are defined by the maximum autoregressive and moving average lags, respectively, with nonzero coefficients. That is, (6.1) must have at least one $\phi_{sp} \neq 0$ and one $\theta_{sq} \neq 0$ over $s = 1, \ldots, S$. Further, as with any ARMA process, $\phi_s(L)$ and $\theta_s(L)$ must have no root in common in order to identify uniquely the parameters of the process; this is extensively discussed in Box and Jenkins (1976).

Periodic ARMA processes have distinctive stationarity and invertibility properties compared with a conventional (constant parameter) ARMA process. As already remarked, the deterministic seasonality considered in Chapter 2 is a special form of a periodic process. This special case of (6.1) has $\phi_s(L)$, $\theta_s(L)$, and $E(\varepsilon_{s\tau}^2)$ constant over $s = 1, \ldots, S$, but the intercept c_s is seasonally varying. It was noted in Chapter 2 that the deterministic seasonal process is not second-order stationary because its mean is dependent on the season. Here, we examine further the distinctive second-order properties of periodic processes by considering three more special cases that illustrate their essential features.

6.2.1 Periodic Heteroskedasticity

Consider initially the constant coefficient special case of (6.1), where

$$\phi(L)y_{s\tau} = c + \theta(L)\varepsilon_{s\tau}, \qquad s = 1, \ldots, S, \ \tau = 1, \ldots, T_\tau, \qquad (6.2)$$

with all roots of $\phi(L)$ and $\theta(L)$ outside the unit circle. This is a conventional ARMA process except that we assume that the disturbance variance $E(\varepsilon_{s\tau}^2) = \sigma_s^2$ is not constant over $s = 1, \ldots, S$. Assuming that any presample starting values have mean μ, we then see that the mean for the process is $E(y_{s\tau}) = \mu = c/\phi(1)$. This mean is unaffected compared with the usual ARMA case, where $E(\varepsilon_{s\tau}^2)$ is constant over s. With seasonal disturbance heteroskedasticity, however, $y_{s\tau} - \mu$ is not second-order stationary.

To illustrate this last point, consider the simple zero-mean seasonal heteroskedastic AR(1) process $y_t = \phi y_{t-1} + \varepsilon_t$ with $|\phi| < 1$. Written in periodic notation, this is

$$y_{s\tau} = \phi y_{s-1,\tau} + \varepsilon_{s\tau}. \qquad (6.3)$$

For simplicity of presentation, we assume that $y_{s\tau}$ corresponds to $s = S$ (the final observation of year τ). Repeated substitution then yields

$$
\begin{aligned}
y_{S\tau} &= \phi^2 y_{S-2,\tau} + \varepsilon_{S\tau} + \phi\varepsilon_{S-1,\tau}, \\
&= \phi^S y_{S,\tau-1} + \varepsilon_{S\tau} + \phi\varepsilon_{S-1,\tau} + \cdots + \phi^{S-1}\varepsilon_{1\tau}, \\
&= \phi^{\tau S} y_{S0} + \sum_{j=0}^{\tau-1} \phi^{Sj}(\varepsilon_{S,\tau-j} + \phi\varepsilon_{S-1,\tau-j} + \cdots + \phi^{S-1}\varepsilon_{1,\tau-j}).
\end{aligned}
$$
$$(6.4)$$

Provided that the single starting value y_{S0} has the same variance as each sample period $y_{S\tau}$, it follows that

$$\text{Var}\,(y_{s\tau}) = \gamma_s(0) = \frac{\sigma_s^2 + \phi^2\sigma_{s-1}^2 + \cdots + \phi^{2(S-1)}\sigma_{s-(S-1)}^2}{1 - \phi^{2S}} \qquad (6.5)$$

for $s = S$. Indeed, (6.5) applies for all $s = 1, \ldots, S$. Thus, $\text{Var}(y_{s\tau})$ is periodically varying since the weighting of each σ_s^2 ($s = 1, \ldots, S$) depends on the season s in which $y_{s\tau}$ is observed. The year τ is, however, irrelevant. For example, assume that $S = 2$, with $\sigma_1^2 = 1$, $\sigma_2^2 = 10$, and $\phi = 0.8$. Then (6.5) implies that $\text{Var}\,(y_{1\tau}) = (1 + 0.8^2 \times 10)/(1 - 0.8^4) = 12.5$ and $\text{Var}\,(y_{2\tau}) = (10 + 0.8^2 \times 1)/(1 - 0.8^4) = 18.0$.

It might also be noted that the process (6.2) has autocovariances at lag k, $\gamma_s(k) = E(y_{s\tau} - \mu)(y_{s-k,\tau} - \mu)$, which are seasonally varying. To emphasize this dependence, the notation $\gamma_s(k)$ for the autocovariance indicates the season

s in addition to the lag k. For example, for (6.3) where $\mu = 0$, multiplying by $y_{s-1,\tau}$ and taking expectations reveals that

$$\gamma_s(1) = E(y_{s\tau} y_{s-1,\tau}) = \phi \gamma_{s-1}(0),$$

which is periodically varying through the variance $\gamma_{s-1}(0)$. (It should be understood that $\gamma_0 = \gamma_S$.) As with the variance, this autocovariance is independent of τ. Thus, the second-order properties themselves follow a seasonal cycle.

At least in principle, the effect of periodic heteroskedasticity in this framework can be easily removed by standardizing by division by the appropriate standard deviation. Specifically, the standardized process $(y_{s\tau} - \mu)/\sqrt{\text{Var}(y_{s\tau})}$ has a zero mean, unit variance, and autocovariances that are independent of s.

6.2.2   Periodic MA(1) Process

This introductory look at the properties of periodic processes now turns to the ARMA coefficients, where we consider the two specific cases of periodic MA(1) and AR(1) processes.

The periodic MA(1) process is

$$y_{s\tau} = \varepsilon_{s\tau} - \theta_s \varepsilon_{s-1,\tau}, \qquad s = 1, \dots, S, \tag{6.6}$$

where, compared with the notation of (6.1), the second subscript on θ has been dropped for notational clarity. To keep the algebra simple, no deterministic component is included in (6.6), so that $y_{s\tau}$ has mean zero, and we assume constant disturbance variance $E(\varepsilon_{s\tau}^2) = \sigma^2$. It is straightforward to see that even without periodic heteroskedasticity through $E(\varepsilon_{s\tau}^2)$, the process $y_{s\tau}$ exhibits periodic heteroskedasticity, since

$$\text{Var}(y_{s\tau}) = \gamma_s(0) = E(\varepsilon_{s\tau} - \theta_s \varepsilon_{s-1,\tau})^2,$$
$$= (1 + \theta_s^2)\sigma^2.$$

Further, the autocovariance at lag 1 is

$$\gamma_s(1) = E(y_{s\tau} y_{s-1,\tau})$$
$$= E(\varepsilon_{s\tau} - \theta_s \varepsilon_{s-1,\tau})(\varepsilon_{s-1,\tau} - \theta_{s-1} \varepsilon_{s-2,\tau})$$
$$= -\theta_s \sigma^2$$

for $s = 1, \dots, S$. Autocovariances $\gamma_s(k)$ for $k > 1$ are all zero. In particular, the autocovariance at lag S, $\gamma_s(S)$ is zero for all s.

Thus, although the periodic MA(1) exhibits periodic variances and autocovariances, observations 1 year apart, $y_{s\tau}$ and $y_{s,\tau-1}$, are not correlated. This implies that the characteristic of seasonality in economic variables that the patterns in the observations tend to repeat each year, and hence that $y_{s,\tau-1}$

provides relevant information for the prediction of $y_{s\tau}$, cannot be delivered by a periodic MA(1) process. This remains true for any periodic MA process of order $S - 1$ or less, and indicates why low-order periodic MA processes have been of little interest in economics.

6.2.3 Periodic AR(1) Process

Now consider the simple periodic AR(1), or PAR(1), process

$$y_{s\tau} = \phi_s y_{s-1,\tau} + \varepsilon_{s\tau}, \qquad s = 1, \dots, S. \tag{6.7}$$

Again with substitution for lagged y, this can be written as

$$
\begin{aligned}
y_{s\tau} &= \phi_s \phi_{s-1} y_{s-2,\tau} + \varepsilon_{s\tau} + \phi_s \varepsilon_{s-1,\tau}, \\
&= \phi_s \phi_{s-1} \cdots \phi_1 y_{s,\tau-1} + \varepsilon_{s\tau} + \phi_s \varepsilon_{s-1,\tau} + \phi_s \phi_{s-1} \varepsilon_{s-2,\tau} \\
&\quad + \cdots + \phi_s \phi_{s-1} \cdots \phi_{s-(S-1)} \varepsilon_{s-(S-1),\tau}.
\end{aligned}
\tag{6.8}
$$

In the writing of (6.8), the cyclic nature of ϕ_s has been exploited and it should be understood that $\phi_i = \phi_{S+i}$ when $i \leq 0$. Notice that by substituting for lagged y precisely S times, the right-hand side of this expression involves the annual lagged value $y_{s,\tau-1}$ together with a periodic MA($S - 1$) process. Irrespective of s, the coefficient of $y_{s,\tau-1}$ is the product of all S periodic AR(1) coefficients, namely $\psi = \phi_1 \phi_2 \cdots \phi_S$. The presence of the periodic moving average process implies that Var($y_{s\tau}$) and its autocovariances vary over s. However, with a starting value such that $E(y_{s0}) = 0$, then the process (6.7) with no intercept has $E(y_{s\tau}) = \mu = 0$.

To illustrate the periodic nature of the second-order properties, consider the quarterly PAR(1) process for $y_{4\tau}$ (again the final observation of year τ is selected for illustrative purposes because this keeps the notation relatively simple). Applying (6.8) and substituting for the annual lagged values $y_{4,\tau-j}$ ($j > 0$), then

$$y_{4\tau} = \psi^\tau y_{40} + \sum_{j=0}^{\tau-1} \psi^j (\varepsilon_{4,\tau-j} + \phi_4 \varepsilon_{3,\tau-j} + \phi_4 \phi_3 \varepsilon_{2,\tau-j} + \phi_4 \phi_3 \phi_2 \varepsilon_{1,\tau-j}).$$

Since the disturbances are uncorrelated, and provided $|\psi| = |\phi_1 \cdots \phi_S| < 1$, then (as $\tau \to \infty$)

$$\text{Var}(y_{4\tau}) = \gamma_4(0) = (1/1 - \psi^2)\left[\sigma_4^2 + \phi_4^2 \sigma_3^2 + \phi_4^2 \phi_3^2 \sigma_2^2 + \phi_4^2 \phi_3^2 \phi_2^2 \sigma_1^2\right].$$

With the use of the same logic, it can similarly be shown that

$$\text{Var}(y_{1\tau}) = \gamma_1(0) = (1/1 - \psi^2)\left[\sigma_1^2 + \phi_1^2 \sigma_4^2 + \phi_1^2 \phi_4^2 \sigma_3^2 + \phi_1^2 \phi_4^2 \phi_3^2 \sigma_2^2\right].$$

Even with homoskedasticity in the disturbances ($\sigma_1^2 = \sigma_2^2 = \sigma_3^2 = \sigma_4^2 = \sigma^2$), the periodic AR(1) process $y_{s\tau}$ exhibits periodic heteroskedasticity. Further, multiplying the right-hand side of (6.7) by $y_{s-1,\tau}$ and taking expectations, the autocovariances at lag 1 for the PAR(1) satisfy

$$\gamma_s(1) = E(y_{s\tau}, y_{s-1,\tau}) = \phi_s \gamma_{s-1}(0), \tag{6.9}$$

which becomes

$$\gamma_4(1) = E(y_{4\tau}, y_{3,\tau}) = \phi_4 \gamma_3(0),$$
$$\gamma_1(1) = E(y_{1\tau}, y_{4,\tau-1}) = \phi_1 \gamma_4(0),$$

when applied with $s = 4$ and $s = 1$, respectively, in the quarterly case.

At the annual lag S, however,

$$\gamma_s(S) = E(y_{s\tau} y_{s,\tau-1}) = \psi \gamma_s(0). \tag{6.10}$$

Notice that while $\gamma_s(1)$ in (6.9) is periodic through both ϕ_s and $\gamma_{s-1}(0)$, $\gamma_s(S)$ is periodic only through the variance $\gamma_s(0)$. Consequently, the autocorrelation of $y_{s\tau}$ at lag S, namely $\rho_s(S) = \gamma_s(S)/\gamma_s(0) = \psi$, is constant over $s = 1, \ldots, S$.

Equation (6.8) also implies that the PAR(1) process gives rise to an annual pattern in the conditional expectations, with

$$E(y_{s\tau} \mid y_{s,\tau-1}) = \psi y_{s,\tau-1},$$

which applies for all s. Thus, in contrast to the periodic MA(1) process, the PAR(1) process gives rise to a type of seasonal habit persistence whereby an annual pattern in the observations will tend to be repeated when ψ is positive.

6.3   Representations and Properties of Periodic Processes

The key to the deeper analysis of periodic processes lies in adoption of a vector representation. Because periodic moving average processes do not appear to be useful in economics (see Subsection 6.2.2), we concentrate on periodic autoregressive processes. Our analysis also allows for seasonal intercept terms and seasonal disturbance heteroskedasticity. After examining the VAR representation in the first subsection, we use this representation to derive the properties of stationary periodic autoregressive processes in the second subsection. The final part of this section examines the theoretical implications of treating a periodic process as a constant coefficient ARMA one.

6.3.1   The VAR Representation

The vector representation has already been exploited for some purposes in Chapters 1 and 2, where the stacked vector of observations for the S seasons of year τ is denoted Y_τ, with U_τ the corresponding vector of disturbances. This representation is especially useful in the periodic case. For example, with quarterly data ($S = 4$) and the addition of seasonal intercepts, the PAR(1) process of (6.7) becomes

$$
\begin{bmatrix} 1 & 0 & 0 & 0 \\ -\phi_2 & 1 & 0 & 0 \\ 0 & -\phi_3 & 1 & 0 \\ 0 & 0 & -\phi_4 & 1 \end{bmatrix} \begin{bmatrix} y_{1\tau} \\ y_{2\tau} \\ y_{3\tau} \\ y_{4\tau} \end{bmatrix}
$$

$$
= \begin{bmatrix} 0 & 0 & 0 & \phi_1 \\ 0 & 0 & 0 & 0 \\ 0 & 0 & 0 & 0 \\ 0 & 0 & 0 & 0 \end{bmatrix} \begin{bmatrix} y_{1,\tau-1} \\ y_{2,\tau-1} \\ y_{3,\tau-1} \\ y_{4,\tau-1} \end{bmatrix} + \begin{bmatrix} c_1 \\ c_2 \\ c_3 \\ c_4 \end{bmatrix} + \begin{bmatrix} \varepsilon_{1\tau} \\ \varepsilon_{2\tau} \\ \varepsilon_{3\tau} \\ \varepsilon_{4\tau} \end{bmatrix}, \tag{6.11}
$$

or, more compactly,

$$
\Phi_0 Y_\tau = \Phi_1 Y_{\tau-1} + C + U_\tau, \tag{6.12}
$$

where $E(U_\tau U_\tau') = \Sigma = \operatorname{diag}(\sigma_1^2, \sigma_2^2, \sigma_3^2, \sigma_4^2)$, while the $S \times S$ coefficient matrices Φ_0 and Φ_1 together with the $S \times 1$ vector of intercepts C are defined by comparison of (6.12) with (6.11).

The representation of the periodic process as a VAR process effectively treats the observations $y_{s\tau}$ for the seasons $s = 1, \ldots, S$ as separate series. Thus, while the representation (6.11) can also be used in the nonperiodic AR(1) case, that case has the cross-equation restrictions $\phi_1 = \phi_2 = \phi_3 = \phi_4 = \phi$ and $\sigma_1^2 = \sigma_2^2 = \sigma_3^2 = \sigma_4^2 = \sigma^2$, but these cross-equation restrictions do not apply in the PAR(1) case. Indeed, given the distinct parameters for each season permitted by the PAR process, the vector representation is natural for this process. It also has the considerable advantage of allowing the exploitation of standard results for the properties of VAR processes, as discussed in, for example, Lütkepohl (1991). The vector representation of a PAR process appears to have been introduced by Gladyshev (1961) and has since been used by many authors, including Franses (1994, 1996b), Osborn (1991), Tiao and Grupe (1980), and Troutman (1979).

The PAR(1) process results in a VAR(1) representation. However, the PAR(p) process

$$
\phi_s(L) y_{s\tau} = c_s + \varepsilon_{s\tau}, \qquad s = 1, \ldots, S, \tag{6.13}
$$

with $\phi_s(L) = 1 - \phi_{s1}L - \cdots - \phi_{sp}L^p$ will, in general, have a VAR(P) representation, where $P = \operatorname{int}[(p + S - 1)/S]$. For example, in the quarterly case ($S = 4$), PAR processes up to and including order 4 have a PAR(1) representation because the fourth-order polynomial $\phi_s(L) = 1 - \phi_{s1}L - \phi_{s2}L^2 - \phi_{s3}L^3 - \phi_{s4}L^4$ can be accommodated in Φ_0 and Φ_1. The precise one-to-one mapping between the polynomial coefficients of the PAR process and the elements of the VAR coefficient matrices is discussed in a number of the papers referred to above. The intuition of this mapping is straightforward, with the sth row of the periodic coefficient matrices containing the coefficients of $\phi_s(L)$.

Thus, the general vector representation for a PAR(p) process is the VAR(P)

$$\Phi_0 Y_\tau = \Phi_1 Y_{\tau-1} + \cdots + \Phi_P Y_{\tau-P} + C + U_\tau \tag{6.14}$$

or, using the matrix polynomial lag operator $\Phi(L) = \Phi_0 - \Phi_1 L^S - \cdots - \Phi_P L^{PS}$,

$$\Phi(L)Y_\tau = C + U_\tau. \tag{6.15}$$

As indicated above, we wish to use the lag operator L as operating on the season, with the annual lag operator being L^S. Therefore, because $Y_{\tau-1}$ is the annual lag of the vector Y_τ, then $L^S Y_\tau = Y_{\tau-1}$ and, in general, $\Phi(L)$ is a matrix polynomial in the annual lag operator. However, a common factor across the $\phi_s(L)$ leads to a nonseasonal factor in $\Phi(L)$ (see Subsection 6.3.3 below), and hence we do not use the notation $\Phi(L^S)$.

Here (6.14) is not written in the standard form of a VAR process, which replaces Φ_0 by an identity matrix and Σ by a general symmetric positive definite matrix. However, the more usual VAR(P) representation can be obtained by inverting Φ_0, so that

$$Y_\tau = \Phi_0^{-1}\Phi_1 Y_{\tau-1} + \cdots + \Phi_0^{-1}\Phi_P Y_{\tau-P} + \Phi_0^{-1}C + \Phi_0^{-1}U_t,$$
$$= A_1 Y_{\tau-1} + \cdots + A_P Y_{\tau-P} + \widetilde{C} + V_\tau, \tag{6.16}$$

where $A_i = \Phi_0^{-1}\Phi_i$ ($i = 1, \ldots, P$), $\widetilde{C} = \Phi_0^{-1}C$, and $V_\tau = \Phi_0^{-1}U_\tau$. It should be noted that $|\Phi_0| = 1$ since this matrix is lower triangular with diagonal elements of unity. Thus it must be nonsingular and the conventional VAR representation (6.16) is well defined.

6.3.2 Properties of Stationary PAR Processes

Stationarity of a VAR system can be examined through the final equation representation, which for (6.14) or equivalently (6.16) is

$$|\Phi(L)| Y_\tau = \Phi^*(1)C + \Phi^*(L)U_\tau, \tag{6.17}$$

where $\Phi^*(L)$ is the adjoint matrix of $\Phi(L)$. The final equation representation has played an important role in the analysis of dynamic econometric models [see, e.g., Wallis (1977)]. Indeed, as discussed in standard analyses of VAR systems, such as Hamilton (1994a), stationarity of the VAR (6.14) requires that all roots of $|\Phi(L)| = 0$ lie outside the unit circle. Since there is a one-to-one mapping between the PAR coefficients of (6.13) and the VAR coefficients of (6.14), then the stationarity condition for the latter also provides the stationarity condition relevant to the PAR(p) process.

To illustrate, we again consider the quarterly PAR(1) process of (6.11), which has

$$|\Phi(L)| = 1 - \phi_1\phi_2\phi_3\phi_4 L^4 \tag{6.18}$$

and

$$
\Phi^*(L) = \begin{bmatrix} 1 & \phi_1\phi_3\phi_4 L^4 & \phi_1\phi_4 L^4 & \phi_1 L^4 \\ \phi_2 & 1 & \phi_1\phi_2\phi_4 L^4 & \phi_1\phi_2 L^4 \\ \phi_2\phi_3 & \phi_3 & 1 & \phi_1\phi_2\phi_3 L^4 \\ \phi_2\phi_3\phi_4 & \phi_3\phi_4 & \phi_4 & 1 \end{bmatrix}, \qquad (6.19)
$$

from which the final equations representation can be explicitly written out for each $y_{s\tau}$. Now, in Subsection 6.2.3 above we discussed certain properties of PAR(1) processes. It can now be verified that the representation (6.8), used a number of times in that analysis, is the final equation representation for a PAR(1) process with $C = 0$.

From the determinantal equation (6.18), it follows that $|\Phi(L)| = 0$ has all roots outside the unit circle if and only if $|\phi_1\phi_2\phi_3\phi_4| < 1$. Therefore, the stationarity condition for the PAR(1) process is $|\phi_1\phi_2\phi_3\phi_4| < 1$. It should also be noted that the condition $|\phi_s| < 1$ $(s = 1, \ldots, S)$ is sufficient but not necessary for stationarity. Therefore, it is permissible for a stationary PAR(1) process to contain one or more individual $|\phi_s| \geq 1$.

Taking expectations in (6.14), (6.16), or (6.17) yields the mean vector for stationary Y_τ as

$$
M = E(Y_\tau) = \Phi(1)^{-1}C = \frac{1}{|\Phi(1)|}\Phi^*(1)C. \qquad (6.20)
$$

Even if the same intercept applies in each equation of the VAR, the mean of the process varies over the seasons because the elements of $\Phi(1)^{-1}$ depend on the periodically varying autoregressive coefficients. Thus, from (6.18) and (6.19), the quarterly PAR(1) can be seen to have mean vector

$$
M = \begin{bmatrix} \mu_1 \\ \mu_2 \\ \mu_3 \\ \mu_4 \end{bmatrix} = \frac{1}{1 - \phi_1\phi_2\phi_3\phi_4} \begin{bmatrix} c_1 + \phi_1\phi_4\phi_3 c_2 + \phi_1\phi_4 c_3 + \phi_1 c_4 \\ c_2 + \phi_2\phi_1\phi_4 c_3 + \phi_2\phi_1 c_4 + \phi_2 c_1 \\ c_3 + \phi_3\phi_2\phi_1 c_4 + \phi_3\phi_2 c_1 + \phi_3 c_2 \\ c_4 + \phi_4\phi_3\phi_2 c_1 + \phi_4\phi_3 c_2 + \phi_4 c_3 \end{bmatrix},
$$

$$(6.21)$$

which is (in general) periodically varying. It might be remarked that a nonperiodic mean is technically possible, but the four elements of (6.21) being equal implies that highly nonlinear restrictions hold involving the intercepts c_s and the original periodic autoregressive coefficients ϕ_s for $s = 1, 2, 3, 4$.

Using standard results for a stationary VAR process described in Lütkepohl (1991), the matrix of autocovariances at lag K, $\Gamma(K) = E(Y_\tau - M)(Y_{\tau-K} - M)'$ for the VAR process (6.16), satisfies the vector Yule–Walker equations

$$
\Gamma(K) = A_1\Gamma(K - 1) + \cdots + A_P\Gamma(K - P), \qquad K \geq P. \qquad (6.22)
$$

As shown by Lütkepohl, the matrices $\Gamma(K - P)$, $K = 1, \ldots, P - 1$ can be uniquely determined as functions of the VAR parameters, namely A_i, $i = 1, \ldots, P$, and $\Omega = E(V_\tau V_\tau')$. In simple cases, however, the required second-order properties of a stationary PAR(p) process can be derived in a more direct way. Indeed, the autocovariance properties for the quarterly PAR(1) have already been illustrated in Subsection 6.2.3 and we will not repeat them here. Nevertheless, it should be emphasized that the lag k used in (6.9) and (6.10) of that discussion is not the same as K of (6.22). This is because the former refers to a single-period lag of $y_{s\tau}$ and the latter to an annual lag of the vector Y_τ. Thus, for the quarterly PAR(1) case, the one-period autocovariances $\gamma_s(1)$ for $s = 2, 3, 4$ are elements of the variance-covariance matrix $\Gamma(0)$, while $\gamma_1(1) = E(y_{1\tau} - \mu_1)(y_{4,\tau-1} - \mu_4)$ is the $(1, 4)$ element of $\Gamma(1)$. It can be verified that the PAR(1) result $\Gamma(1) = A\Gamma(0)$ implies $\gamma_1(1) = \phi_1 \gamma_4(0)$, in accordance with (6.9).

Viewed as a stationary VAR process, estimation and inference for a PAR(p) process becomes straightforward. Indeed, because the disturbances of the PAR are uncorrelated, then (with known starting values for the process) the maximum likelihood estimation is equivalent to OLS applied separately for each season. Similarly, standard t and F distribution tests are asymptotically valid. However, restrictions across the seasons $s = 1, \ldots, S$ of the PAR process would imply restrictions across the equations of the VAR(P), and imposition of these restrictions requires simultaneous estimation of the S equations of the PAR.

Lütkepohl (1991) discusses the application of various tests to the stationary periodic model to investigate the nature of the seasonal variation found in observed values for a time series. In particular, the general PAR(p) model can be tested against restricted versions to examine whether all parameters vary with the seasons, whether there is periodic heteroskedasticity, whether only the intercept varies seasonally with all other parameters constant, and so on. Likelihood ratio (LR) tests are convenient for this purpose. Under the appropriate null hypothesis, all LR tests in this context follow asymptotic χ^2 distributions, with degrees of freedom equal to the number of restrictions.

6.3.3   The Constant Parameter Representation of a PAR Process

Following Tiao and Grupe (1980), Osborn (1991) studies the properties of a PAR process when it is analyzed as an ARMA process. That is, these authors examine the theoretical properties that ensue when a periodic process is incorrectly analyzed as a constant parameter univariate process. In order to focus on the stochastic part of the periodic process, we assume the zero-mean process with $C = 0$.

As illustrated for the PAR(1) process in Subsection 6.2.3, the autocovariances of a periodic autoregressive process are seasonally varying. However, for a stationary periodic process we can define an underlying theoretical autocovariance at lag k as

$$
\begin{aligned}
\gamma(k) &= \lim_{T \to \infty} \frac{1}{T} \sum_{t=1}^{T} y_t y_{t-k}, \\
&= \frac{1}{S} \sum_{s=1}^{S} \lim_{T \to \infty} \frac{1}{T_\tau} \sum_{\tau=1}^{T_\tau} y_{s\tau} y_{s-k,\tau}, \\
&= \frac{1}{S} \sum_{s=1}^{S} \gamma_s(k), \qquad k = 0, 1, \ldots.
\end{aligned}
\tag{6.23}
$$

Thus, if the true periodic variation is ignored, then the sample autocovariance computed using observations over all seasons $s = 1, \ldots, S$ will asymptotically tend to $\gamma(k)$ that is equal to the average of the S underlying periodic autocovariances. Tiao and Grupe (1980) investigate the nature of the univariate ARMA process that has autocovariances $\gamma(k)$, $k = 0, 1, \ldots$, while Osborn (1991) goes further in linking this implied ARMA process to that which results when component ARMA processes are aggregated.

The existence of this unique univariate representation with autocovariances $\gamma(k)$ for $k = 0, 1, \ldots$ implies that a standard univariate analysis will never uncover the presence of an underlying periodic DGP. In that sense the periodic process is described as "hidden" by Tiao and Grupe (1980). Aside from understanding the true nature of the time series relationship, recognition of the periodic nature of the DGP should also, at least theoretically, confer improved forecast accuracy [Tiao and Grupe (1980), Osborn (1991)].

To explore the nature of the misspecified univariate representation of a periodic process, recall that aggregation of component ARMA processes produces an ARMA process; papers on this topic include those by Lütkepohl (1984), Rose (1977), and Wei (1978). As illustrated by the analysis of Rose, the aggregation of such processes effectively proceeds by expressing the component processes in terms of a common autoregressive polynomial, which (in general) becomes the autoregressive component of the ARMA process for the aggregate. Osborn (1991) shows that the autoregressive component $|\Phi(L)|$ of the final equation system (6.17) becomes the autoregressive component in the univariate ARMA representation, while the moving average order of the final equation system is the maximum order implied by $\Phi^*(L)U_\tau$. As already noted, $|\Phi(L)|$ is (in general) a seasonal AR, while the MA order has to take account of the intrayear lags implied by $\Phi^*(L)U_\tau$. In the first subsection below we illustrate this analysis for the case of a quarterly PAR(1) process; in the second subsection we examine how this generalizes to higher-order cases.

The Quarterly PAR(1) Process

From the discussion of the immediately preceding subsection, the final equation representation for a zero-mean quarterly PAR(1) process is

$$(1 - \phi_1\phi_2\phi_3\phi_4 L^4) \begin{bmatrix} y_{1\tau} \\ y_{2\tau} \\ y_{3\tau} \\ y_{4\tau} \end{bmatrix}$$

$$= \begin{bmatrix} \varepsilon_{1\tau} + \phi_1\varepsilon_{4,\tau-1} + \phi_1\phi_4\varepsilon_{3,\tau-1} + \phi_1\phi_3\phi_4\varepsilon_{2,\tau-1} \\ \varepsilon_{2\tau} + \phi_2\varepsilon_{1\tau} + \phi_1\phi_2\varepsilon_{4,\tau-1} + \phi_1\phi_2\phi_4\varepsilon_{3,\tau-1} \\ \varepsilon_{3\tau} + \phi_3\varepsilon_{2\tau} + \phi_2\phi_3\varepsilon_{1\tau} + \phi_1\phi_2\phi_3\varepsilon_{4,\tau-1} \\ \varepsilon_{4\tau} + \phi_4\varepsilon_{3\tau} + \phi_3\phi_4\varepsilon_{2\tau} + \phi_2\phi_3\phi_4\varepsilon_{1\tau} \end{bmatrix}. \qquad (6.24)$$

The common autoregressive polynomial of the final equations, namely $(1 - \phi_1\phi_2\phi_3\phi_4 L^4) = 1 - \psi L^4$, provides the autoregressive component of the constant parameter representation. Since the final equation for each $y_{s\tau}$ involves disturbances with lags of up to three quarters, then the MA component is of order 3. Thus, in the notation of Box and Jenkins (1976), the misspecified constant parameter representation is a SARMA(0,3) × (1,0)$_4$ process.

Osborn (1991) further analyzes the nature of this moving average component. For the quarterly PAR(1) process, this is equivalent to the moving average that results from aggregating the four hypothetical MA(3) processes

$$(1 + \phi_2 L + \phi_2\phi_3 L^2 + \phi_2\phi_3\phi_4 L^3)v_{1\tau},$$
$$(1 + \phi_3 L + \phi_3\phi_4 L^2 + \phi_1\phi_3\phi_4 L^3)v_{2\tau},$$
$$(1 + \phi_4 L + \phi_1\phi_4 L^2 + \phi_1\phi_2\phi_4 L^3)v_{3\tau},$$
$$(1 + \phi_1 L + \phi_1\phi_2 L^2 + \phi_1\phi_2\phi_3 L^3)v_{4\tau},$$

that have disturbances $v_{s\tau}$ independent over s and τ and with $E(v_{s\tau}^2) = \sigma_s^2/4$ for $s = 1, \ldots, 4$. Notice that the moving average term in the sth equation here collects the moving average coefficients for the disturbance $\varepsilon_{s\tau}$ across the four equations of (6.24). Combined with standard results for the aggregation of moving average processes, the analysis enables the parameters of the misspecified univariate representation to be derived.

This discussion generalizes in a fairly intuitive way to PAR(1) processes for data observed at other frequencies. A monthly PAR(1) process has, for example, a constant parameter SARMA(0,11) × (1,0)$_{12}$ representation. The discussion also applies to the nonstationary PAR processes examined in the next section.

Nevertheless, the implied univariate representation is of a relatively complicated form. The simplest realistic case is the quarterly PAR(1) with univariate SARMA(0,3) × (1,0)$_4$ representation, with higher-order processes having more complex univariate representations. With the emphasis within economics on using autoregressive models in practice, this suggests that an autoregressive model selected by conventional criteria will be an approximation to the underlying univariate representation. The combination of a seasonal autoregressive component (from the final equation representation) and a nonseasonal one arising from an autoregressive approximation to the MA may lead to a high-order AR model being required to mimic, with a reasonable level of accuracy, the autocorrelation properties that arise from the true periodic DGP.

Higher-Order PAR Processes

Some care is required when considering PAR processes of order greater than 1. A genuinely periodic autoregressive process requires that the coefficient polynomials $\phi_s(L)$ are not the same over $s = 1, \ldots, S$. This does not, however, prevent these polynomials containing one or more factors that are common to all S polynomials. For example, the PAR(2) case with a common factor $(1 - \alpha L)$ is

$$(1 - \phi_{s1}L - \phi_{s2}L^2) = (1 - \beta_s L)(1 - \alpha L), \qquad s = 1, \ldots, S.$$
$$(6.25)$$

If there is such a common factor in the periodic autoregressive polynomials, then this may be taken out first and the above analysis applied to $(1 - \alpha L)y_{s\tau}$. This implies that, for example, a quarterly PAR(2) process with a common autoregressive factor $(1 - \alpha L)$ has a constant parameter SARMA(1,3) $\times$ (1,0)$_4$ representation.

The analysis for general PAR processes is more complicated. Nevertheless, results from the PAR(1) case can be exploited if the PAR(p) is seen to arise as the product of p PAR(1) processes; this approach is used by Franses (1996b). To illustrate it, consider the PAR(2) process

$$y_{s\tau} - \alpha_s y_{s-1,\tau} = \beta_s(y_{s-1,\tau} - \alpha_{s-1}y_{s-2,\tau}) + \varepsilon_{s\tau}, \qquad s = 1, \ldots, S,$$
$$(6.26)$$

which is a PAR(1) process in the variable $(y_{s\tau} - \alpha_s y_{s-1,\tau})$. In terms of the polynomial coefficient $\phi_s(L)$, this implies

$$\phi_s(L) = 1 - \phi_{s1}L - \phi_{s2}L^2 = 1 - (\beta_s + \alpha_s)L + \beta_s\alpha_{s-1}L^2. \quad (6.27)$$

The coefficients β_s and α_s of (6.26) can be used to define matrices $B_1(L)$ and $B_0(L)$, respectively, in a manner analogous to the definition of $\Phi(L)$ for the PAR(1). The vector representation then becomes

$$B_1(L)B_0(L)Y_\tau = U_\tau,$$
$$(6.28)$$

with the final equation representation

$$|B_1(L)| \, |B_0(L)|Y_\tau = B_0^*(L)B_1^*(L)U_\tau.$$
$$(6.29)$$

Here the asterisk again indicates the adjoint matrix. From (6.29), the nature of the misspecified constant parameter ARMA process can be deduced by using the results for the first-order case. For example, with quarterly data, (6.18) indicates that the common autoregressive polynomial for the PAR(2) is, in general, the SAR(2) one $(1 - \beta_1\beta_2\beta_3\beta_4 L^4)(1 - \alpha_1\alpha_2\alpha_3\alpha_4 L^4)$. Further, (6.19) implies that the MA order for each $y_{s\tau}$ is anticipated to be 6. Thus, the constant parameter representation of the quarterly PAR(2) process is, in general, a SARMA$(0,6) \times (2,0)_4$.

There is no tension between the discussion here and the case of a PAR(2) with a common factor $(1 - \alpha L)$. The common factor would imply that each $\alpha_s = \alpha$ in the matrix $B_0(L)$. This, in turn, leads to cancellation across each equation of (6.29) so that the final equation becomes

$$(1 - \alpha L)|B_1(L)|Y_\tau = B_1^*(L)U_\tau,$$

which is equivalent to applying the PAR(1) analysis to $(1 - \alpha L)y_t$. The possibility of such a common factor and the implied cancellations in (6.29) is covered by our qualification that the results apply in general. Another specific case of interest occurs when not all the individual $\phi_s(L)$ polynomials are of order 2. One or more of these being of order 1 within an overall PAR(2) would imply that the constant parameter representation has a SAR(1) component rather than SAR(2), although the moving average order of 6 may remain unaffected.

The approach embodied in (6.28) can be generalized to handle higher-order PAR processes, with the vector representation for a PAR(p) having the form

$$B_{p-1}(L) \cdots B_1(L)B_0(L)Y_\tau = U_\tau. \tag{6.30}$$

Provided that all the PAR(p) coefficients ϕ_{si} ($s = 1, \ldots, S, i = 1, \ldots, p$) are nonzero, then (in general) the constant parameter process has a seasonal autoregressive component of order p and a moving average component that may be of a very high order.

6.4   Nonstationary Univariate PAR Processes

As discussed already (Subsection 6.3.2), the stationarity condition for the PAR process is that the roots of the determinantal polynomial $|\Phi(L)|$ lie outside the unit circle. First-order unit root nonstationarity arises when this polynomial contains either the seasonal factor $1 - L^S$ or nonseasonal factor $1 - L$, with all other roots being of modulus greater than one. However, in the PAR context, three distinct types of integrated processes give rise to such first-order unit root nonstationarity. These are:

- $y_t \sim I(1)$. This arises when each periodic autoregressive operator $\phi_s(L)$ contains the common factor $\Delta_1 = (1 - L)$, but the matrix representation for $\Delta_1 y_{s\tau}$ is a stationary VAR process.

- $y_t \sim SI(1)$. This arises when each periodic autoregressive operator $\phi_s(L)$ contains the common factor $\Delta_S = (1 - L^S)$, with the matrix representation for $\Delta_S y_{s\tau}$ being a stationary VAR process.
- $y_t \sim PI(1)$. This arises when $|\Phi(L)|$ contains the factor $(1 - L^S)$, but Δ_S is not common to each polynomial $\phi_s(L)$ ($s = 1, \ldots, S$), with the VAR for $\Delta_S y_{s\tau}$ being stationary. This case is termed periodic integration by Osborn et al. (1988).

The first two cases imply stationary PAR processes in the appropriately differenced variable, namely $\Delta_1 y_t$ or $\Delta_S y_t$ respectively, while the third is a specific type of integration that can arise only in the periodic context. The first subsection below discusses these possible forms of integration in a PAR process, and this is followed by an examination of the role of intercepts and trends in nonstationary periodic processes. Strategies for testing for the presence of the usual seasonal unit roots in periodic processes are considered by Ghysels, Hall, and Lee (1996), as discussed in Subsection 6.4.3 below. Through taking a cointegration approach, Subsection 6.4.4 draws together periodic integration and the seasonal unit root tests of HEGY. Subsection 6.4.5 then examines a test explicitly designed for periodic integration. All these initial discussions are simplified by excluding deterministic components and augmentation, with the implications of including such terms considered in the final subsection.

6.4.1 Types of Integration in a PAR Process

We will consider first the simple quarterly PAR(1) process of (6.11) with $C = 0$. First-order periodic integration, or $y_t \sim PI(1)$, arises when the process is nonstationary with $\phi_1 \phi_2 \phi_3 \phi_4 = 1$, but not all individual $\phi_s = 1$. For stationarity to be induced in this process, the quasi-differences $D_s y_{s\tau} = y_{s\tau} - \phi_s y_{s-1,\tau}$ are required. Note that we specify not all $\phi_s = 1$ to rule out the possibility that $y_{s\tau} \sim I(1)$. Put in a slightly different way, a conventionally integrated $I(1)$ process can be viewed as a special case of a $PI(1)$ process with $\phi_s = 1$ for $s = 1, \ldots, S$. To distinguish periodic integration, however, this special case is ruled out.

Recall again that the stationarity condition in the quarterly PAR(1) process arises from the polynomial $1 - \phi_1 \phi_2 \phi_3 \phi_4 L^4$ of its final equation representation (6.24). Thus, unit root nonstationarity $\phi_1 \phi_2 \phi_3 \phi_4 = 1$ gives rise to $\Delta_4 = 1 - L^4$ as the autoregressive operator in the final equations. It is, therefore, tempting to conclude that $y_{s\tau}$ is seasonally integrated in the sense defined in Chapter 3, since it appears that seasonal differencing is required to render the process stationary. This conclusion is, however, false.[1] Seasonal differencing imposes the autoregressive operator $1 - L^4$ on each $y_{s\tau}$ and hence forces the transformed

---

[1] Franses (1996b) contains a similar discussion to that here.

variables to have the form of the final equation representation of the PAR process. That is, the final equation representation then becomes more than simply a statistical tool; rather it captures the form of the observed variables after transformation. The final equation representation for the PAR(1) case has the vector moving average component $\Phi^*(L)U_\tau$, with $\Phi^*(L)$ shown for $S = 4$ in (6.19). Putting $L^4 = 1$ in the latter matrix, and using the identity $\phi_4 = 1/(\phi_1\phi_2\phi_3)$ implied by periodic integration for this process, yields

$$\Phi^*(1) = \begin{bmatrix} 1 & 1/\phi_2 & 1/\phi_2\phi_3 & \phi_1 \\ \phi_2 & 1 & 1/\phi_3 & \phi_1\phi_2 \\ \phi_2\phi_3 & \phi_3 & 1 & \phi_1\phi_2\phi_3 \\ 1/\phi_1 & 1/\phi_1\phi_2 & 1/\phi_1\phi_2\phi_3 & 1 \end{bmatrix}. \tag{6.31}$$

It is straightforward to see that this matrix has rank one, since any row is a multiple of any other row, and hence it is singular. The vector moving average process of $\Delta_4 Y_\tau = \Phi^*(L)U_\tau$ is, consequently, noninvertible [Lütkepohl (1991)]. Therefore, applying Δ_4 represents overdifferencing and the process is not seasonally integrated. Nor, indeed, is the process of a conventional $I(1)$ form. Since $|\Phi(L)|$ contains the factor Δ_4, first differencing is not sufficient to remove the nonstationarity present in $y_{s\tau}$.

The same problem does not arise when quasi-differences $D_s y_{s\tau} = y_{s\tau} - \phi_s y_{s-1,\tau}$ are taken. In this special PAR(1) case, the quasi-differences yield $D_s y_{s\tau} = \varepsilon_{s\tau}$, where the $\varepsilon_{s\tau}$ are uncorrelated. More generally, however, if the PAR(p) process of (6.13) is $PI(1)$, then the quasi-difference is a stationary process that may be periodic. To examine the more general quarterly $PI(1)$ process, we can use the approach of Subsection 6.3.3. Thus, in particular, for $y_{s\tau}$ to be $PI(1)$, one of the matrices in (6.30) must be nonstationary with determinant $(1 - L^4)$. Taking this to be the matrix $B_0(L)$, then $\alpha_1\alpha_2\alpha_3\alpha_4 = 1$ and the quasi-difference defined by

$$D_s y_{s\tau} = y_{s\tau} - \alpha_s y_{s-1,\tau} \tag{6.32}$$

will be a stationary process. In the PAR(2) case, for example, this quasi-difference is given by (6.26), and this quasi-difference series is a stationary PAR(1) process provided $|\beta_1\beta_2\beta_3\beta_4| < 1$.

The conclusion from this discussion is that a $PI(1)$ process is neither $I(1)$ nor $SI(1)$. Therefore, the three possibilities distinguished above for the types of integration possible in a stochastic nonstationary periodic process are, indeed, distinct. Further, the quasi-differenced $PI(1)$ process will generally be periodic. Nevertheless, it is possible that the $PI(1)$ process could be nonperiodic after the periodic nonstationarity is removed. On the other hand, if a periodic autoregressive process is $I(1)$ or $SI(1)$, then the stationary autoregressive process after appropriate differencing (Δ_1 or Δ_S respectively) must be periodic.

To formalize this discussion and generalize it to any number of seasons S, we can define first-order periodic integration as follows.

Definition 6.1: The nonstationary periodic process $y_{s\tau}$ is said to be periodically integrated of order 1 if there exist quasi-differences $D_s y_{s\tau} = y_{s\tau} - \alpha_s y_{s-1,\tau}$, with $\prod_{s=1}^{S} \alpha_s = 1$ and not all $\alpha_s = 1$, such that the VAR representation for the quasi-differences is stationary and invertible.

6.4.2 The Role of Intercept and Trend Terms

Deterministic terms have important implications in nonstationary periodic processes. To begin with the periodically integrated case, note that the PAR process with an intercept but no trend for each equation has the final equation representation (6.17). For a periodically integrated PAR(1) process, this becomes

$$\Delta_S Y_\tau = \Phi^*(1)C + \Phi^*(L)U_\tau$$

and hence

$$E(\Delta_S Y_\tau) = \Phi^*(1)C, \qquad (6.33)$$

so that the implied mean annual growth is itself seasonally varying. This is true whether or not the same intercept applies in each equation, since the periodic autoregressive coefficients influence the mean growth for each season through the matrix $\Phi^*(1)$. To see this, note that using $\Phi^*(1)$ as given for the quarterly case in (6.31) yields

$$\Phi^*(1)C = \begin{bmatrix} c_1 + c_2/\phi_2 + c_3/\phi_2\phi_3 + c_4\phi_1 \\ c_1\phi_2 + c_2 + c_3/\phi_3 + c_4\phi_1\phi_2 \\ c_1\phi_2\phi_3 + c_2\phi_3 + c_3 + c_4\phi_1\phi_2\phi_3 \\ c_1/\phi_1 + c_2/\phi_1\phi_2 + c_3/\phi_1\phi_2\phi_3 + c_4 \end{bmatrix}.$$

Irrespective of the values of the intercepts $c_1, \ldots, c_4$, periodic integration implies that the rows of $\Phi^*(1)C$ are multiples of each other, with ϕ_1, ϕ_2, ϕ_3 providing the scaling factors (recall in this periodic integration case that $\phi_4 = 1/\phi_1\phi_2\phi_3$). Thus, there are only two ways that the four elements of $\Phi^*(1)C$ here can be equal. The first case is that in which no equation includes an intercept, that is $C = 0$, and hence $E(\Delta_4 Y_\tau) = 0$. Obviously this rules out any underlying trend in $E(y_{s\tau})$. The second possibility is that $\phi_1 = \phi_2 = \phi_3 = \phi_4$, but this is ruled out by the periodic nature of the process. The arguments generalize to higher-order PAR processes that are periodically integrated and any $PI(1)$ process with nonzero intercepts gives rise to a seasonally varying trend in $E(y_{s\tau})$. A common deterministic trend can arise only in the trivial case where this trend has zero slope.

The discussion of the previous paragraph is built on the more detailed analysis of Paap and Franses (1999), who also allow the $PI(1)$ process to include (linear) trend terms in the equations for quarterly $y_{s\tau}$. With no restrictions, such (nonzero) trends imply seasonally varying quadratic trends in $E(y_{s\tau})$ because $E(\Delta_4 y_{s\tau})$ is generally a linear function of τ. However, nonlinear restrictions can be applied to the trend coefficients so that $E(\Delta_4 y_{s\tau})$ is not a function of τ and hence no quadratic trends enter. Further, Paap and Franses establish that the inclusion of trends allows the possibility of a nonzero linear trend in $E(y_{s\tau})$ that is constant over s. Such a constant linear deterministic trend also requires nonlinear restrictions to apply across the coefficients of the equations for $y_{s\tau}$ ($s = 1, \ldots, 4$); the authors examine these restrictions.

For a periodic trend stationary process, Paap and Franses show that a common trend similarly requires nonlinear restrictions to hold across the four separate equations for $y_{s\tau}$. These restrictions are fewer than those of the $PI(1)$ case. Another case of interest is when the periodic process $y_{s\tau} \sim SI(1)$, so that $\Delta_4 y_{s\tau}$ is stationary. Although not examined by Paap and Franses, the same principles apply. Nonzero intercepts will give rise to trends, which generally vary over the seasons (see Chapter 3). For $E(y_{s\tau})$ to have a common linear trend, $E(\Delta_4 y_{s\tau})$ must be constant over season s and year τ. Clearly, this requires no trend in the equations for the stationary $\Delta_4 y_{s\tau}$ and a mean $E(\Delta_4 y_{s\tau})$ which is constant over $s = 1, \ldots, 4$. As noted in Subsection 6.3.2 above, a constant mean for a stationary periodic variable requires restrictions to hold across the equations for the four seasons. Thus, restrictions have to be imposed across the equations for $\Delta_4 y_{s\tau}$ to yield a common trend in $E(y_{s\tau})$.

Finally, we might consider the case in which periodic $y_{s\tau}$ is $I(1)$. While a nonzero intercept in the equation for $\Delta_1 y_{s\tau}$ will give rise to a trend in $E(y_{s\tau})$, this trend is constant over seasons. This arises because considered over a year, $\Delta_4 y_{s\tau} = \Delta_1 y_{s\tau} + \cdots + \Delta_1 y_{s+1,\tau-1}$. Hence the expected annual change is always equal to the expected one-period changes summed over all four quarters, and so the slope of the annual trend in $E(y_{s\tau})$ is invariant to the season. Nonzero trends are not typically included, since these can give rise to quadratic trends in $E(y_{s\tau})$. For trends to be included and a common trend to arise, restrictions need to be imposed on the trend coefficients in order that $E(\Delta_4 y_{s\tau})$ is constant over both s and τ.

Our discussion of earlier chapters has noted that an underlying trend in $E(y_{s\tau})$ that is constant over seasons is often considered to be plausible for economic data. To summarize the discussion of this subsection, imposing such a constant trend on a nonstationary periodic process requires a different specification of the deterministic components for each type of possible nonstationarity in $y_{s\tau}$. Specifically, taking the inclusion of intercepts for granted:

- $y_{s\tau} \sim I(0)$. Trends must be included in the equations for $y_{s\tau}$, with restrictions across equations for $s = 1, \ldots, 4$.

- $y_{s\tau} \sim I(1)$. No trends are generally included, with no other restrictions required. If trends are included, restrictions across equations are required.
- $y_{s\tau} \sim SI(1)$. No trends are included; restrictions are required across equations.
- $y_{s\tau} \sim PI(1)$. Trends are included, with restrictions across equations.

Although we have concentrated on the quarterly case, the essential arguments generalize to any S.

6.4.3 Testing for Seasonal Integration in a PAR Process

Ghysels, Hall, and Lee (1996), or GHL, propose using the HEGY test of Chapter 3 to examine the null hypothesis $y_t \sim SI(1)$ in the context of a possibly periodic autoregressive process. GHL discuss separate tests for each of the real unit roots $+1$, -1 and for each complex pair of unit roots implied by Δ_S, in addition to joint tests of the overall null hypothesis. Here we set out the joint test and indicate the nature of the asymptotic distributions under the null hypothesis. We also consider how GHL approach testing the $I(1)$ null hypothesis. To keep the issues as simple as possible, we continue to base the discussion on the DGP of a (possibly seasonal) random walk with no drift and zero starting values. In this subsection, we also assume homoskedasticity across quarters in the disturbance variance. Thus, most of our discussion is conducted in the context of a nonperiodic DGP. Nevertheless, the GHL approach allows the possibility that the underlying process may be periodic. The discussion later in the chapter makes some of the relevant periodic issues clearer.

The test regression of GHL generalizes the HEGY test regression by allowing the possibility that each coefficient could vary over the seasons of the year. Thus, in the quarterly context the GHL test regression for testing a $SI(1)$ null hypothesis is, with no deterministic components or augmentation,

$$\Delta_4 y_t = \sum_{s=1}^{4} \pi_{s1} \delta_{st} y_{t-1}^{(1)} - \sum_{s=1}^{4} \pi_{s2} \delta_{st} y_{t-1}^{(2)}$$

$$- \sum_{s=1}^{4} \pi_{s3} \delta_{st} y_{t-2}^{(3)} - \sum_{s=1}^{4} \pi_{s4} \delta_{st} y_{t-1}^{(3)} + \varepsilon_t, \tag{6.34}$$

where δ_{st} is the quarterly dummy variable that takes the value one when t falls in season s and is zero otherwise and $y_t^{(1)}$, $y_t^{(2)}$, and $y_t^{(3)}$ are the HEGY transformed variables discussed at length in Chapter 3.

As previously emphasized, the HEGY variables $y_{t-1}^{(1)}$, $y_{t-1}^{(2)}$, $y_{t-2}^{(3)}$, and $y_{t-1}^{(3)}$ are asymptotically mutually orthogonal by construction. However, this asymptotic orthogonality does not apply to the season-specific variables $\delta_{st} y_{t-1}^{(1)}$, $\delta_{st} y_{t-1}^{(2)}$, $\delta_{st} y_{t-2}^{(3)}$ and $\delta_{st} y_{t-1}^{(3)}$ for given s. Essentially, the orthogonality holds when all s

are considered because off-setting effects apply across the values of s. Restriction to a specific s rules out such off-setting effects. Nevertheless, from the definition of seasonal dummy variables, the two sets of variables $(\delta_{st} y_{t-1}^{(1)}, \delta_{st} y_{t-1}^{(2)}, \delta_{st} y_{t-2}^{(3)}, \delta_{st} y_{t-1}^{(3)})$ and $(\delta_{qt} y_{t-1}^{(1)}, \delta_{qt} y_{t-1}^{(2)}, \delta_{qt} y_{t-2}^{(3)}, \delta_{qt} y_{t-1}^{(3)})$ must be mutually orthogonal for $s \neq q$. This implies some simplification for the scaled matrix of regressors, $T_\tau^{-2} X'X$, with each of the 4×4 submatrices for (6.34) being diagonal. Notice, incidentally, that the scaling we use involves T_τ, the number of observations available for each quarterly series $y_{s\tau}$, rather than the total number of observations T. The former is appropriate since the regressors in (6.34) effectively involve only observations for a single quarter.

The seasonal integration null hypothesis implies $\pi_{si} = 0 \, (s = 1, \ldots, 4, i = 1, \ldots, 4)$. GHL adopt a Wald test statistic that is, when the joint zero coefficient null hypothesis is true, equal to

$$
\begin{aligned}
W(\widehat{\pi}) &= \widehat{\pi}'(X'X)\widehat{\pi}/\widetilde{\sigma}^2, \\
&= U'X(X'X)^{-1}X'U/\widetilde{\sigma}^2, \\
&= (1/\widetilde{\sigma}^2)(U'X/T_\tau)\left(X'X/T_\tau^2\right)^{-1}(X'U/T_\tau),
\end{aligned}
$$

where $\widehat{\pi} = (\widehat{\pi}_{11}, \ldots, \widehat{\pi}_{41}, \widehat{\pi}_{12}, \ldots, \widehat{\pi}_{44})'$ is the vector OLS coefficient estimator in (6.34), X is the $(T \times 16)$ regressor matrix, U is the $(T \times 1)$ complete sample vector of disturbances $U' = (\varepsilon_1, \ldots, \varepsilon_T)$, and $\widetilde{\sigma}^2$ is the usual OLS estimator of the disturbance variance. Although the form of $T_\tau^{-2} X'X$ allows some simplification in the derivation of the asymptotic distributions, the simple summation of distributions obtained for the HEGY test of Chapter 3 does not apply here. For our present purposes, further details are unnecessary.

In the context of regression (6.34) with and without deterministic terms, GHL present critical values for testing the $SI(1)$ null hypothesis $\pi_{si} = 0, s = 1, \ldots, 4, i = 1, \ldots, 4$. Further, to jointly test the seasonal unit roots $-1, \pm i$, critical values are supplied for a test of $\pi_{si} = 0, s = 1, \ldots, 4, i = 2, 3, 4$. Analogous values are also presented for the monthly case. Nevertheless, it should be remarked that a practical disadvantage of the overall GHL test is that it is highly parameterized. This $SI(1)$ test requires the examination of 16 coefficients for quarterly data, with monthly data using $12^2 = 144$ coefficients.

Due to the orthogonality of the seasonal dummy variables, the asymptotic distribution associated with the GHL test statistic does have a simpler form when a periodic approach is taken to testing the unit root for a specific frequency. For example, to examine the null hypothesis of an $I(1)$ process, GHL consider the test regression

$$
\Delta y_t = \sum_{s=1}^{4} \pi_{s1} \delta_{st} y_{t-1} + \varepsilon_t
$$

where the null hypothesis is $\pi_{s1} = 0, s = 1. \ldots, 4$. In this case, the scaled regressor matrix $T_\tau^{-2} X'X$ is diagonal. Its four diagonal elements can be written in the double subscript notation as $T_\tau^{-2} \sum_{\tau=1}^{T_\tau} [y_{4,\tau-1}]^2$, $T_\tau^{-2} \sum_{\tau=1}^{T_\tau} [y_{1\tau}]^2$, $T_\tau^{-2} \sum_{\tau=1}^{T_\tau}$ $[y_{2\tau}]^2$ and $T_\tau^{-2} \sum_{\tau=1}^{T_\tau} [y_{3\tau}]^2$ respectively. Note that each regressor here is observed once a year. For example, with $s = 1, \delta_{1t} y_{t-1}$, is nonzero only when t falls in the first quarter of year τ and then

$$\delta_{1t} y_{t-1} = y_{4,\tau-1}$$
$$= y_{4,\tau-2} + \varepsilon_{1,\tau-1} + \varepsilon_{2,\tau-1} + \varepsilon_{3,\tau-1} + \varepsilon_{4,\tau-1}$$
$$= \sum_{q=1}^{4} \sum_{j=1}^{\tau-1} \varepsilon_{qj}.$$

The implication is that the sequence of nonzero values for $\delta_{1t} y_{t-1}$ behaves like the sum of four random walk processes, with one process associated with each of the seasons of the year. This type of behavior is familiar from the discussion of Chapter 3. The same logic can be applied for any s, with $\delta_{st} y_{t-1}$ exhibiting this behavior.

To illustrate the derivation of the asymptotic distributions for this last test regression, we will consider a single "t-ratio", $t(\widehat{\pi}_{s1})$. Under the null hypothesis $\pi_{s1} = 0$ we have (continuing to use the double subscript notation)

$$t(\widehat{\pi}_{s1}) = \frac{T_\tau^{-1} \sum_{\tau=1}^{T_\tau} \varepsilon_{s\tau} y_{s-1,\tau}}{\widetilde{\sigma} [T_\tau^{-2} \sum_{\tau=1}^{T_\tau} (y_{s-1,\tau})^2]^{1/2}}$$
$$\Rightarrow \frac{\int_0^1 [\sum_{i=1}^4 W_i(r)] dW_s(r)}{[\int_0^1 \{\sum_{i=1}^4 W_i(r)\}^2 dr]^{1/2}}, \tag{6.35}$$

where the second line uses the standard results for Brownian motions summarized in Chapter 3.[2] The results also derive from the properties just discussed for $\delta_{st} y_{t-1}$. Note, in particular, that unlike the result for $t(\widehat{\pi}_1)$ in the original HEGY test, the distribution in (6.35) is not a Dickey–Fuller distribution. The essential reason is that although the regressor $\delta_{st} y_{t-1}$ effectively involves the sum of random walks which can be associated with the four quarters, the relevant disturbance term $\varepsilon_{s\tau}$ in (6.35) relates to only one of these quarters (namely, the quarter s in which observation t falls). In contrast, by imposing a common coefficient π_1, the usual HEGY regression does not separate the quarters and hence the disturbances for all quarters are included when the asymptotic distribution of $t(\widehat{\pi}_1)$ is considered.

Because of the mutual orthogonality of the regressors in the $I(1)$ test regression of GHL, the distribution of their overall Wald test statistic for the

[2] Here we also implicitly assume that the starting values for the process $y_{s0} = 0$ ($s = 1, \ldots, 4$).

quarterly case is obtained as the simple sum of the squares of the distributions in (6.35) for $s = 1, 2, 3, 4$. GHL provide these quarterly critical values, and also those appropriate for the monthly case. As usual, these are obtained by simulation.

6.4.4 Cointegration Approaches To Testing Seasonal Integration

Dickey et al. (1984) and Osborn (1993) observe that if y_t is a seasonal random walk process, then the S processes obtained by considering $y_{s\tau}$ separately for $s = 1, \ldots, S$ are independent random walk processes. As noted by Osborn, and discussed in Chapter 3 above, this implies that the $y_{s\tau}$ processes cannot be cointegrated with each other. Franses (1994) adopts this idea and develops a cointegration approach for testing the null hypothesis of seasonal integration. As we also demonstrate below, cointegration is also implicit in the GHL test just considered.

The Multivariate Test of Franses

Even if the underlying process may be a constant parameter one, Franses (1994) advocates treating the series for the S seasons separately and adopting a vector representation. A univariate process up to order S has a VAR(1) representation, $\Phi_0 Y_\tau = \Phi_1 Y_{\tau-1} + U_\tau$ as in (6.12). Assuming zero starting values and no drift terms, this VAR(1) can also be written as

$$\Delta_S Y_\tau = A Y_{\tau-1} + V_\tau, \tag{6.36}$$

where $A = -(I_S - \Phi_0^{-1}\Phi_1)$ and $V_\tau = \Phi_0^{-1} U_\tau$. Considered as a test regression, it can be noted that (6.36) is a vector version of the DHF regression of Chapter 3.

Franses applies results of Johansen (1988) relating to cointegration in systems of equations to this vector representation. Specifically, the seasonal random walk implies $A = 0$ and $V_\tau = U_\tau$. Clearly, the rank of A is then zero. Within the cointegration framework, this zero rank implies that the S processes $y_{s\tau}$ ($s = 1, \ldots, S$) are individually $I(1)$ and not cointegrated. When the rank of A is denoted as r, then a test of the no cointegration null hypothesis, $r = 0$, is also a test that the univariate process $y_t \sim SI(1)$.

As shown by Johansen (1988), his trace statistic for testing the null hypothesis that $A = 0$ asymptotically has the distribution

$$\left[\int_0^1 W(r)\,dW(r)' \right]' \left[\int_0^1 W(r)W(r)'\,dr \right]^{-1} \left[\int_0^1 W(r)\,dW(r)' \right],$$

$$\tag{6.37}$$

where $W(r) = [W_1(r), \ldots, W_S(r)]'$ is the S-dimensional vector of standard Brownian motions that has sth element associated with the random walk $y_{s\tau}$ for season s. These Brownian motions $W_s(r)$ are the same as those just used in the discussion of the GHL test.

Equation (6.37) is related to the square of the DHF distribution, with these being identical in the special case of $S = 1$ (the Dickey–Fuller case). However, the Franses method is highly parameterized compared with the DHF approach and the other seasonal unit root tests considered in Chapter 3. In the simplest case, in which the seasonal random walk $\Delta_S y_t = \varepsilon_t$ is the true DGP, the cointegration approach implies the estimation of an unrestricted $S \times S$ matrix A, whereas the DHF test involves only a single parameter while the HEGY approach leads to estimation of S parameters ($\pi_1, \ldots, \pi_4$ in the quarterly case). It is, therefore, not surprising that Franses (1994) finds that his method lacks power compared with the HEGY approach.

When cointegration is considered in a VAR system, the Johansen approach is usually implemented as a sequential procedure; see, for example, Banerjee, Dolado, Galbraith, and Hendry (1993). That is, the null hypothesis $r = 0$ (here implying a seasonally integrated process) is tested against the alternative $r > 0$. If this null is rejected, then the null hypothesis $r = 1$ is tested against $r > 1$. This procedure continues until the null hypothesis cannot be rejected and this specific value, say r^*, is taken as the rank of A. Franses (1994) shows that different ranks and restrictions for A are associated with alternative hypotheses of interest in the context of seasonality.

In particular, $y_t \sim PI(1)$ implies that A in the quarterly case has rank $r = 3$. For the simple PAR(1) case, this can be seen from

$$A = -\left(I_4 - \Phi_0^{-1}\Phi_1\right) = \begin{bmatrix} -1 & 0 & 0 & \phi_1 \\ 0 & -1 & 0 & \phi_1\phi_2 \\ 0 & 0 & -1 & \phi_1\phi_2\phi_3 \\ 0 & 0 & 0 & \phi_1\phi_2\phi_3\phi_4 - 1 \end{bmatrix},$$

which results from the matrices of (6.11). Clearly, the final row of this matrix contains only zeros when $\phi_1\phi_2\phi_3\phi_4 = 1$, but the remaining three rows are linearly independent. In this case, the three cointegrating relationships[3] can be defined as $y_{2\tau} - \phi_2 y_{1\tau}$, $y_{3\tau} - \phi_3 y_{2\tau}$, and $y_{4\tau} - \phi_4 y_{3\tau}$, which correspond to three of the quasi-differences discussed above. The fourth quasi-difference, $y_{1\tau} - \phi_1 y_{4,\tau-1}$, also defines a cointegrating relationship, but this is not linearly independent of the others since it is implied by the other three. This carries over to higher-order quarterly PAR processes that are periodically integrated. The quasi-differences of (6.32) for such a $PI(1)$ process define stationary linear combinations of the nonstationary $y_{s\tau}$. Once again, three of the

---

[3] The three cointegrating relationships can be defined by any three of the four quasi-differences.

quasi-differences define linearly independent cointegrating relations, with the fourth implied by these three. The initial insight that periodic integration must imply cointegration between the nonstationary quarterly series $y_{s\tau}$ appears to be from Osborn (1993).

Another case giving rise to rank $r = 3$ occurs when $y_t \sim I(1)$ but contains no seasonal unit roots. In this situation each of pair of variables $(y_{2\tau}, y_{1\tau})$, $(y_{3\tau}, y_{2\tau})$, and $(y_{4\tau}, y_{3\tau})$ is again cointegrated, but now the stationary linear combinations can be defined as $y_{2\tau} - y_{1\tau}$, $y_{3\tau} - y_{2\tau}$, and $y_{4\tau} - y_{3\tau}$. Analogous to the $PI(1)$ case, the cointegrating relationship $y_{1\tau} - y_{4,\tau-1}$ is implied by the other three. Clearly, this $I(1)$ case implies the restrictions that the coefficients of the cointegrating relationships are $(1, -1)$; these restrictions should be tested after establishing that the rank of A is 3. The Johansen methodology [Johansen (1988, 1991), Johansen and Juselius (1990)] can be used to conduct this test, as discussed by Franses (1994). Thus, in this framework, the distinction between a $PI(1)$ process and one that is $I(1)$ is the validity (or not) of these restrictions. Another possibility is that the VAR process for the vector Y_τ contains the three seasonal unit roots -1 and $\pm i$, but not the zero-frequency unit root of 1. This implies $r = 1$ with the unique cointegrating vector being $y_{1\tau} + y_{2\tau} + y_{3\tau} + y_{4\tau}$. Indeed, in the context of a VAR(1) for the quarters, Franses (1994) explicitly examines the nature of the restrictions that must apply for the process y_t to contain any (real) combination of the roots examined by Hylleberg et al. (1990).

The GHL Test in a Vector Context

A different perspective on the GHL test of Subsection 6.4.3 can be given by considering this test in the framework of a VAR. As argued above, a VAR is the natural framework for analyzing periodic processes. Since the alternative hypothesis of the GHL test allows for periodic coefficients, the argument applies here also.

In order to write the GHL test regression (6.34) in vector form, note first that from the definitions of Subsection 3.2.4 the quarterly HEGY variables satisfy

$$
\begin{aligned}
y_t^{(1)} &= y_{t-1}^{(1)} + \Delta_4 y_t, \\
y_t^{(2)} &= -y_{t-1}^{(2)} + \Delta_4 y_t, \\
y_t^{(3)} &= -y_{t-2}^{(3)} + \Delta_4 y_t.
\end{aligned}
\tag{6.38}
$$

Now consider the VAR equation implied by the GHL test regression (6.34) for the second quarter ($s = 2$), which is

$$
\begin{aligned}
\Delta_4 y_{2\tau} &= \pi_{21} y_{1\tau}^{(1)} - \pi_{22} y_{1\tau}^{(2)} - \pi_{23} y_{4,\tau-1}^{(3)} - \pi_{24} y_{1\tau}^{(3)} + \varepsilon_{2\tau}, \\
&= \pi_{21}\left(y_{4,\tau-1}^{(1)} + \Delta_4 y_{1\tau}\right) - \pi_{22}\left(-y_{4,\tau-1}^{(2)} + \Delta_4 y_{1\tau}\right) \\
&\quad - \pi_{23} y_{4,\tau-1}^{(3)} - \pi_{24}\left(-y_{3,\tau-1}^{(3)} + \Delta_4 y_{1\tau}\right) + \varepsilon_{2\tau}.
\end{aligned}
$$

The latter expression involves the same regressors as those that apply in a HEGY regression for $s = 1$ (namely $y_{4,\tau-1}^{(1)}$, $y_{4,\tau-1}^{(2)}$, $y_{3,\tau-1}^{(3)}$ and $y_{4,\tau-1}^{(3)}$), together with $\Delta_4 y_{1\tau}$, which is stationary under the null hypothesis. Similarly, by repeatedly using the relationships of (6.38), the regressors relevant for $s = 3$ and $s = 4$ can also be written in terms of $y_{4,\tau-1}^{(1)}$, $y_{4,\tau-1}^{(2)}$, $y_{3,\tau-1}^{(3)}$, and $y_{4,\tau-1}^{(3)}$, together with $\Delta_4 y_{1\tau}$, $\Delta_4 y_{2\tau}$ and (for $s = 4$ only) $\Delta_4 y_{3\tau}$. This yields the vector representation of (6.34) as

$$
\begin{bmatrix} \Delta_4 y_{1\tau} \\ \Delta_4 y_{2\tau} \\ \Delta_4 y_{3\tau} \\ \Delta_4 y_{4\tau} \end{bmatrix} = \begin{bmatrix} \pi_{11} & -\pi_{12} & -\pi_{13} & -\pi_{14} \\ \pi_{21} & \pi_{22} & \pi_{24} & -\pi_{23} \\ \pi_{31} & -\pi_{32} & \pi_{33} & \pi_{34} \\ \pi_{41} & \pi_{42} & -\pi_{44} & \pi_{43} \end{bmatrix} \begin{bmatrix} y_{4,\tau-1}^{(1)} \\ y_{4,\tau-1}^{(2)} \\ y_{3,\tau-1}^{(3)} \\ y_{4,\tau-1}^{(3)} \end{bmatrix}
$$

$$
+ \begin{bmatrix} 0 & 0 & 0 & 0 \\ \phi_{21} & 0 & 0 & 0 \\ \phi_{31} & \phi_{32} & 0 & 0 \\ \phi_{41} & \phi_{42} & \phi_{43} & 0 \end{bmatrix} \begin{bmatrix} \Delta_4 y_{1\tau} \\ \Delta_4 y_{2\tau} \\ \Delta_4 y_{3\tau} \\ \Delta_4 y_{4\tau} \end{bmatrix} + \begin{bmatrix} \varepsilon_{1\tau} \\ \varepsilon_{2\tau} \\ \varepsilon_{3\tau} \\ \varepsilon_{4\tau} \end{bmatrix}, \quad (6.39)
$$

where the nonzero coefficients ϕ_{sk} are linear functions of the elements π_{sj} ($j = 1, \ldots, 4$). Taking the right-hand side terms in $\Delta_4 y_{1\tau}$, $\Delta_4 y_{2\tau}$, and $\Delta_4 y_{3\tau}$ to the left-hand side and defining the vector $Y_{\tau-1}^{(H)} = (y_{4,\tau-1}^{(1)}, y_{4,\tau-1}^{(2)}, y_{3,\tau-1}^{(3)}, y_{4,\tau-1}^{(3)})'$, we can write this as the matrix equation

$$
\Phi_0 \Delta_4 Y_\tau = \tilde{\Pi} Y_{\tau-1}^{(H)} + U_\tau,
$$

where the elements of $\tilde{\Pi}$ are the appropriately signed π_{sj} from (6.39). In this vector context, we can also allow periodic heteroskedasticity, so that $E(\varepsilon_{s\tau}^2) = \sigma_s^2$.

Further, again using the HEGY definitions of Chapter 3 we have

$$
Y_{\tau-1}^{(H)} = \begin{bmatrix} y_{4,\tau-1}^{(1)} \\ y_{4,\tau-1}^{(2)} \\ y_{3,\tau-1}^{(3)} \\ y_{4,\tau-1}^{(3)} \end{bmatrix} = \begin{bmatrix} 1 & 1 & 1 & 1 \\ -1 & 1 & -1 & 1 \\ -1 & 0 & 1 & 0 \\ 0 & -1 & 0 & 1 \end{bmatrix} \begin{bmatrix} y_{1,\tau-1} \\ y_{2,\tau-1} \\ y_{3,\tau-1} \\ y_{4,\tau-1} \end{bmatrix} = H Y_{\tau-1}.
$$

Substituting this into the previous equation, our VAR version of the GHL test has been reparameterized as

$$
\Phi_0 \Delta_4 Y_\tau = \tilde{\Pi} H Y_{\tau-1} + U_\tau.
$$

Inverting Φ_0, we have

$$
\Delta_4 Y_t = A Y_{\tau-1} + \Phi_0^{-1} U_\tau, \quad (6.40)
$$

where, in this case, $A = \Phi_0^{-1} \tilde{\Pi} H$. Under the seasonal integration null hypothesis, all $\pi_{sj} = 0$, implying $\tilde{\Pi} = 0$, $\Phi_0 = I_4$, and $A = 0$. Indeed, because Φ_0 and H are nonsingular, the rank of A must always be the same as that of $\tilde{\Pi}$.

Written as in (6.40), it is clear that the overall GHL test can be viewed essentially as a reparameterization of the Franses cointegration test of (6.36). This is analogous to the relationship between the HEGY and Kunst tests for seasonal integration noted in Chapter 3. The fundamental distinction between the Franses and GHL tests is the nature of the hypotheses that are of interest under the alternative, with the Franses approach able to consider periodic integration while the GHL test allows the separate roots of Δ_4 to be examined season by season.

6.4.5 Testing Periodic and Nonperiodic Integration

Although the Franses multivariate test is able to consider periodic integration, the highly parameterized nature of that test is a practical disadvantage. Further, the Franses test effectively begins by considering the null hypothesis that $y_t \sim SI(1)$. Boswijk and Franses (1996) develop a test specifically for the null hypothesis $y_t \sim PI(1)$. Once again, this test is developed in the context of quarterly series. This test exploits the $PI(1)$ implication that the matrix A in the VAR representation (6.36) has a cointegrating rank of 3 and, therefore, that the series $y_{1\tau}$, $y_{2\tau}$, $y_{3\tau}$, and $y_{4\tau}$ possess a single common stochastic trend.

This single common stochastic trend can be illustrated by the case of the periodically quarterly integrated PAR(1) process as written in the final equation representation of (6.24). Repeated here for clarity, this representation is

$$\left(1 - \phi_1\phi_2\phi_3\phi_4 L^4\right) \begin{bmatrix} y_{1\tau} \\ y_{2\tau} \\ y_{3\tau} \\ y_{4\tau} \end{bmatrix}$$

$$= \begin{bmatrix} \varepsilon_{1\tau} + \phi_1\varepsilon_{4,\tau-1} + \phi_1\phi_4\varepsilon_{3,\tau-1} + \phi_1\phi_3\phi_4\varepsilon_{2,\tau-1} \\ \varepsilon_{2\tau} + \phi_2\varepsilon_{1\tau} + \phi_1\phi_2\varepsilon_{4,\tau-1} + \phi_1\phi_2\phi_4\varepsilon_{3,\tau-1} \\ \varepsilon_{3\tau} + \phi_3\varepsilon_{2\tau} + \phi_2\phi_3\varepsilon_{1\tau} + \phi_1\phi_2\phi_3\varepsilon_{4,\tau-1} \\ \varepsilon_{4\tau} + \phi_4\varepsilon_{3\tau} + \phi_3\phi_4\varepsilon_{2\tau} + \phi_2\phi_3\phi_4\varepsilon_{1\tau} \end{bmatrix}.$$

We can then define a disturbance series relating to $y_{1\tau}$ as

$$\eta_\tau = \varepsilon_{1\tau} + \phi_1\varepsilon_{4,\tau-1} + \phi_1\phi_4\varepsilon_{3,\tau-1} + \phi_1\phi_3\phi_4\varepsilon_{2,\tau-1}.$$

Since $y_t \sim PI(1)$ implies $\phi_1\phi_2\phi_3\phi_4 = 1$, the final equation representation for $y_{1\tau}$ can be written as $y_{1\tau} = y_{1,\tau-1} + \eta_\tau$. By repeated substitution in this equation and assuming zero starting values, we have

$$y_{1\tau} = \sum_{j=1}^{\tau} \eta_j.$$

Thus, $\sum \eta_j$ can be defined as the stochastic trend of $y_{1\tau}$. Now, $y_{2\tau} = \varepsilon_{2\tau} + \phi_2 y_{1\tau}$, $y_{3\tau} = \varepsilon_{3\tau} + \phi_3 \varepsilon_{2\tau} + \phi_2 \phi_3 y_{1\tau}$, and $y_{4\tau} = \varepsilon_{4\tau} + \phi_4 \varepsilon_{3\tau} + \phi_3 \phi_4 \varepsilon_{2\tau} + \phi_2 \phi_3 \phi_4 y_{1\tau}$, so that these series contain the stochastic trends $\phi_2 \sum \eta_j$, $\phi_2 \phi_3 \sum \eta_j$, and $\phi_2 \phi_3 \phi_4 \sum \eta_j$, respectively, implying that $\sum \eta_j$ is common to $y_{1\tau}$, $y_{2\tau}$, $y_{3\tau}$, and $y_{4\tau}$. In this $PI(1)$ case the four Brownian motions underlying (appropriately scaled) $y_{s\tau}$ as $\tau \to \infty$ can all be written in terms of a univariate standard Brownian motion $W_\eta(r)$, which in turn derives from $\sum \eta_j$.

In the absence of any augmentation and deterministic terms, the test regression proposed by Boswijk and Franses (1996) has the form

$$y_t = \sum_{s=1}^{4} \phi_s \delta_{st} y_{t-1} + \varepsilon_t. \tag{6.41}$$

Under the null hypothesis, nonlinear least-squares estimation is required in order to impose the periodic integration restriction $\phi_1 \phi_2 \phi_3 \phi_4 = 1$, while linear least squares is appropriate under the alternative hypothesis that $\phi_1 \phi_2 \phi_3 \phi_4 \neq 1$. A conventional likelihood ratio test is used to compare the results of these two estimations and, under the null hypothesis, the authors show that asymptotically this test statistic converges to

$$LR \Rightarrow \frac{\left[\int_0^1 W_\eta(r) \, dW_\eta(r) \right]^2}{\int_0^1 [W_\eta(r)]^2 \, dr}, \tag{6.42}$$

which is the square of the Dickey–Fuller t-distribution. Although we omit the technical details in this case, the intuition is clear. A $PI(1)$ process contains precisely one underlying unit root, represented here by the stochastic trend $\Sigma \eta_j$, and a test of this unit root is effectively a Dickey–Fuller test.

Boswijk and Franses also consider testing the null hypothesis of constant autoregressive coefficients over the quarters, which is a null hypothesis that the DGP is not periodic. In a quarterly first-order autoregressive model that may exhibit periodically varying coefficients, this implies testing the null hypothesis $\phi_1 = \phi_2 = \phi_3 = \phi_4$. When the presence of a unit root is also suspected, the Boswijk–Franses periodic unit root test may be applied first, followed by a test of constant autoregressive coefficients, or the tests may be applied in the opposite sequence. In either case, Boswijk and Franses establish that for testing the constant coefficients null hypothesis, a conventional χ^2 test is valid asymptotically, whether the series contain a unit root or not.

A different strategy is to conduct the unit root and constant coefficient tests jointly. Such a test can be obtained through the GHL approach by using the test regression

$$\Delta y_t = \sum_{s=1}^{4} \pi_{s1} \delta_{st} y_{t-1} + \varepsilon_t \tag{6.43}$$

with the null hypothesis $\pi_{s1} = 0$, $s = 1, 2, 3, 4$ (recall that the coefficients π_{s1} test the conventional unit root of 1 for each of the seasons). From the results of Subsection 6.4.3, and especially (6.35), the Wald test statistic for this null hypothesis is

$$W(\widehat{\pi}_{11} \cap \widehat{\pi}_{21} \cap \widehat{\pi}_{31} \cap \widehat{\pi}_{41}) \Rightarrow \sum_{s=1}^{4} \frac{\left\{ \int_0^1 \left[\sum_{i=1}^4 W_i(r) \right] dW_s(r) \right\}^2}{\int_0^1 \left\{ \sum_{i=1}^4 W_i(r) \right\}^2 dr}.$$

Boswijk and Franses show that this expression is the sum of two independent variables, which are a χ^2 variable with three degrees of freedom (which can be associated with a test of constant autoregressive coefficients) and the square of a Dickey–Fuller t distribution (associated with the test of a unit root).

6.4.6 *Extensions: Augmentation and Deterministic Components*

As with the seasonal integration tests considered in Chapter 3, the test regressions for examining nonstationarity in periodic autoregressive processes can be augmented with additional lags without affecting the substantive asymptotic results. Naturally, however, the augmentation here should generally be of a periodic type. For the Franses and GHL tests in which the overall null hypothesis is $y_t \sim SI(1)$, the augmentation is with additional lags of $\Delta_4 y_{s\tau}$ that are stationary under the null hypothesis, and hence conventional significance tests are valid. The Boswijk and Franses test has $y_t \sim PI(1)$ under the null hypothesis, and augmentation is with additional lags of the quasi-difference $D_s y_{s\tau}$. Conventional significance tests are, however, valid here too since under the null hypothesis $D_s y_{s\tau}$ is a stationary variable.

Augmentation by lagged $D_s y_{s\tau}$ in the Boswijk and Franses approach is equivalent to augmentation by lagged levels of y_t. To see this, recall the second-order PAR(2) process of (6.26), namely

$$y_{s\tau} - \alpha_s y_{s-1,\tau} = \beta_s(y_{s-1,\tau} - \alpha_{s-1} y_{s-2,\tau}) + \varepsilon_{s\tau}.$$

This becomes the first-order process $y_{s\tau} = \alpha_s y_{s-1,\tau} + \varepsilon_{s\tau}$ when $\beta_s = 0$. Hence augmentation of the equation for $D_s y_{s\tau} = y_{s\tau} - \alpha_s y_{s-1,\tau}$ by the lagged quasi-difference $y_{s-1,\tau} - \alpha_{s-1} y_{s-2,\tau}$ can be reparameterized as augmenting the equation for $y_{s\tau}$ with the additional term $y_{s-2,\tau}$. Even without the imposition of the periodic integration restriction $\alpha_1 \cdots \alpha_S = 1$, we would anticipate that hypothesis tests relating to these augmentations in the two parameterizations will be (at least asymptotically) equivalent, and hence that the order of augmentation can be determined by using conventional test statistics in the levels representation. It must be emphasized, however, that this argument relates to the order of augmentation beyond the first-order lag in the equation for $y_{s\tau}$. Obviously, under the periodic integration null hypothesis, the distribution of the estimated coefficient $\widehat{\alpha}_s$ in the PAR(1) case is nonstandard.

Turning to the issue of the deterministic components to be included in the test regressions, the discussion of Subsection 6.4.2 provides relevant background. Nevertheless, the issue of what deterministic components should be included in a unit root test regression is a slightly different one to the examination of the implications of such terms within the true DGP. As usual, of course, the inclusion of constant or seasonally varying deterministic terms generally affects the critical values even if they have zero coefficients in the DGP.

The GHL test takes seasonal integration as its overall null hypothesis, and as in Chapter 3, seasonally varying intercepts should be included in the test regression to allow for nonzero starting values for the process. Further, a periodic $SI(1)$ process with drift will generally imply seasonally varying means of $\Delta_S y_{s\tau}$. To deliver invariance of the test statistics to possible periodic growth rates, the logic of Kiviet and Phillips (1992) implies that seasonally varying trend coefficients should also be included. This also allows for the possibility under the alternative hypothesis that the $y_{s\tau}$ are stationary around seasonally varying linear trends. The tables of GHL consider this case. However, this is an overparameterization if the researcher wishes to consider an alternative hypothesis involving a trend that is constant over seasons, since this would imply restrictions in the test regression (see the discussion of Subsection 6.4.2). The extension to allow the GHL approach to be undertaken in a regression with restrictions of this type is not examined by GHL or other authors, but it may be worthy of further investigation.

Similar reasoning applies to the Franses cointegration test, especially as we argue above that the GHL test and the Franses test are effectively the same. Once again, the general recommendation is to include intercept and trend terms in each of the equations of the VAR test regression. However, for the reasons just discussed in the GHL case, unit root testing in an appropriately restricted VAR model might be considered.

One obvious possibility is to begin with no restrictions relating to the deterministic terms and to conduct a pretest of the validity of the desired restrictions prior to testing the unit roots. Such testing is not straightforward, however, because the asymptotic distributions of the test statistics relating to the deterministic terms generally depend on whether the unit root restrictions are valid or not. Another possibility is to consider a joint test that relates to the nature of $E(y_{s\tau})$ and the presence of unit roots. Paap and Franses (1999) use this strategy in the context of the Boswijk–Franses periodic integration test. At the time of writing, however, it is unclear whether a joint testing approach is to be preferred to a periodic unit root test with the inclusion of unrestricted periodic intercepts and trend terms in the test regression.

To date, no investigation appears to have been undertaken in which the test regression is specified with restrictions relating to the deterministic terms imposed a priori on the basis of the plausibility of their implications. In particular, the use of a nonseasonal trend is common when testing for seasonal unit roots in a nonperiodic context, on the grounds that seasonally varying trends are typically implausible for economic series. On the same basis, the restrictions required to deliver a nonseasonal trend might also be considered in the context of unit root testing for periodic processes.

6.5   Periodic Cointegration

In an analogous way to cointegration between two nonseasonal $I(1)$ time series y_t, x_t yielding a stationary linear combination $y_t - kx_t$, periodic cointegration can be defined as the existence of the stationary linear combinations $y_{s\tau} - k_s^P x_{s\tau}$ ($s = 1, \ldots, S$) between two integrated series. The superscript P here denotes that the parameter is periodically varying; this superscript is used so that we can later distinguish periodic and seasonal cointegration. The study of Birchenhall et al. (1989) was the first to consider periodic cointegration, with subsequent developments being undertaken primarily by Franses and his co-authors [including Boswijk and Franses (1995), Franses and Kloek (1995), Kleibergen and Franses (1999)].

This section discusses some aspects of periodic cointegration and error correction. Other overviews are provided by Franses (1996a) and Franses and McAleer (1998). To keep the analysis as simple as possible, we restrict our attention to the case in which cointegration is considered for two quarterly variables, y_t and x_t.

6.5.1   Some Issues in Periodic Cointegration

Based on Boswijk and Franses (1995), we can define periodic cointegration for our case as follows:

Definition 6.2: Consider the eight annually observed variables $y_{s\tau}, x_{s\tau}$ for $s = 1, \ldots, 4$, each of which is $I(1)$. The variables x and y are fully periodically cointegrated of order (1,1) if there exist coefficients k_s^P such that $y_{s\tau} - k_s^P x_{s\tau}$ is stationary for $s = 1, \ldots, 4$ with not all k_s^P equal. The variables are partially cointegrated if k_s^P exist such that $y_{s\tau} - k_s^P x_{s\tau}$ is stationary for only some $s = 1, \ldots, 4$.

By considering the quarterly series $y_{s\tau}, x_{s\tau}$ for $s = 1, \ldots, 4$ as separate series of annual observations, this definition allows the possibility that each of the time series y_t, x_t could be $I(1), SI(1)$, or $PI(1)$. This follows Boswijk and Franses. Unlike the definition of these authors, however, our definition does not permit the cointegrating vector to be common across all four quarters.

Thus, our definition is genuinely periodic in the sense that the cointegrating relationship is periodically varying. This is in the spirit of the distinction drawn in Section 6.4 between a $PI(1)$ process and a conventional $I(1)$ process.

Despite the apparent flexibility of allowing the two time series to be $I(1)$, $SI(1)$, or $PI(1)$, some possibilities are ruled out in the context of periodic cointegration. Thus, if y_t, $x_t \sim I(1)$ with $y_{s\tau} - kx_{s\tau} \sim I(0)$ for some quarter s, then it logically follows that the cointegrating vector for all $s = 1, \ldots, 4$ must be the nonperiodic one $(1, -k)$. This can be seen by considering $\Delta_1 y_{s\tau} = v_{s\tau}^y$ and $\Delta_1 x_{s\tau} = v_{s\tau}^x$ (where $v_{s\tau}^y$ and $v_{s\tau}^x$ are stationary processes) together with $y_{s\tau} - kx_{s\tau} = y_{s,\tau-1} - kx_{s,\tau-1} + v_{s,\tau}^y - kv_{s\tau}^x$. Since the left-hand side of this expression is stationary, so must be the right-hand side, implying that y_{t-1} and x_{t-1} are cointegrated with cointegrating vector $(1, -k)$. Thus, two $I(1)$ processes cannot be periodically cointegrated; they are either nonperiodically cointegrated for all $s = 1, \ldots, 4$ or not cointegrated for any s.

With the use of a similar approach, it can be seen that if two $PI(1)$ processes are cointegrated for one quarter, then they must be cointegrated for all quarters. That is, periodically integrated processes must be either fully cointegrated or not cointegrated at all; they cannot be partially cointegrated. Consequently, the conclusion reached by Franses and Paap (1995b) that German consumption and income are both $PI(1)$ but only cointegrated in the first quarter of the year is logically impossible. To explore the nature of cointegration a little further, note that the $PI(1)$ properties imply that $y_{s\tau} = \alpha_s^y y_{s-1,\tau} + v_{s\tau}^y$ and $x_{s\tau} = \alpha_s^x x_{s-1,\tau} + v_{s\tau}^x$, with $v_{s\tau}^y$, $v_{s\tau}^x$ again stationary and with $\alpha_1^i \alpha_2^i \alpha_3^i \alpha_4^i = 1(i = y, x)$. If $y_{s\tau} - k_s^P x_{s\tau}$ is stationary, then (using the same logic as in the preceding paragraph) so must be $\alpha_s^y y_{s-1,\tau} - k_s^P \alpha_s^x x_{s-1,\tau}$. The cointegrating vector between $y_{s-1,\tau}$, $x_{s-1,\tau}$ is $(1, -k_{s-1}^P)$, where $k_{s-1}^P = k_s^P \alpha_s^x / \alpha_s^y$, and hence the cointegration is periodic unless $\alpha_s^x = \alpha_s^y$ for all $s = 1, \ldots, 4$.

It is, however, possible for two $SI(1)$ processes to be periodically cointegrated, either fully or partially. This is because the four quarterly series in each case (namely $y_{s\tau}$ and $x_{s\tau}$ for $s = 1, \ldots, 4$) follow separate unit root processes and hence cointegration may occur between any or all of the $(y_{s\tau}, x_{s\tau})$ pairs.[4] Thus, periodic cointegration can be viewed as an alternative route to be explored to that of seasonal cointegration (examined in Chapter 3).

A slightly different perspective comes from considering x_t to be exogenous with given unit root properties. Then, with the assumption that the cointegrating relationship(s) reflects underlying economic relationships, the nature of the nonstationarity in y_t will be determined by that of x_t together with cointegration. We pursue this approach in the remainder of this section. From this perspective, there are three possibilities to consider:

---

[4] Indeed, it is possible to have cointegration between $y_{s\tau}$ and $x_{j\tau}$ ($j \neq s$), but we rule out this case for simplicity of exposition.

1. $x_t \sim I(1)$. With full periodic cointegration between y_t and x_t, then $y_t \sim PI(1)$. This follows algebraically since we assume that

$$y_{s\tau} = k_s^P x_{s\tau} + u_{s\tau}, \qquad (6.44)$$

with $u_{s\tau}$ stationary. This, with $x_t \sim I(1)$, implies

$$y_{s\tau} = k_s^P x_{s-1,\tau} + k_s^P v_{s\tau}^x + u_{s\tau}.$$

Now, periodic cointegration also implies

$$y_{s-1,\tau} = k_{s-1}^P x_{s-1,\tau} + u_{s-1,\tau},$$

with $u_{s,\tau-1}$ stationary. Therefore, multiplying the last equation by k_s^P / k_{s-1}^P and subtracting from the preceding one, we obtain

$$y_{s\tau} - \frac{k_s^P}{k_{s-1}^P} y_{s-1,\tau} = k_s^P v_{s\tau}^x + u_{s\tau} - \frac{k_s^P}{k_{s-1}^P} u_{s-1,\tau}, \qquad (6.45)$$

where all right-hand side variables are stationary. Therefore, the left-hand side is also stationary. Since the coefficient of $y_{s-1,\tau}$ is $\alpha_s^y = k_s^P / k_{s-1}^P$ (with $k_0^P = k_4^P$), the $PI(1)$ restriction $\alpha_1^y \alpha_2^y \alpha_3^y \alpha_4^y = 1$ must hold for y_t. Thus, for example, a univariate $PI(1)$ process for consumption (y_t here) may be a consequence of an $I(1)$ process for income (x_t) combining with periodic cointegration arising from seasonally varying preferences; see Osborn (1988) for a consumption model based on seasonal preferences. In the special case of nonperiodic cointegration with $k_1^P = \cdots = k_4^P$, then y_t is also $I(1)$ and periodic integration does not result.

2. $x_t \sim PI(1)$. With either full periodic cointegration or nonperiodic cointegration, then in general $y_t \sim PI(1)$ also. With periodic cointegration, however, the coefficients from the cointegrating relationships are confounded with those from the $PI(1)$ process for x_t, so that y_t has distinctive PI coefficients from those of x_t. In particular, following the same logic as that of case 1, instead of (6.45) we obtain

$$y_{s\tau} - \left(k_s^P \alpha_s^x / k_{s-1}^P \right) y_{s-1,\tau} = k_s^P v_{s\tau}^x + u_{s\tau} - \left(k_s^P \alpha_s^x / k_{s-1}^P \right) u_{s-1,\tau}$$

and hence $\alpha_s^y = (k_s^P \alpha_s^x / k_{s-1}^P)$. It can be verified that the $PI(1)$ restriction $\alpha_1^y \alpha_2^y \alpha_3^y \alpha_4^y = 1$ must hold for y_t in this case also. If the cointegration is nonperiodic, then $\alpha_s^y = \alpha_s^x$ and hence x and y have identical periodic properties.

3. $x_t \sim SI(1)$ combined with periodic cointegration between x_t and y_t. This case has effectively been considered above. The four unit root processes in $x_{s\tau}$ ($s = 1, 2, 3, 4$) give rise to four distinct unit root processes in y_t through $y_{s\tau} = k_s^P x_{s\tau} + u_{s\tau}$. Therefore, $y_t \sim SI(1)$. Note that the $SI(1)$ conclusion

for y_t carries over also in the special case of nonperiodic cointegration between y_t and x_t. Notice, in particular, that y_t cannot be either $I(1)$ or $PI(1)$ when periodic or nonperiodic integration exists between $y_{s\tau}$ and $x_{s\tau}$; such cointegration cannot reduce the number of unit roots from the four unit roots in the $SI(1)$ process for x_t.

The different implications of these cases for the unit root properties of y_t imply, in turn, that no single ECM representation is appropriate for all cases of periodic cointegration. More particularly, periodic cointegration implies that the ECM appropriate for case 1 has the form

$$D_s y_{s\tau} = \lambda_s^P (y_{s-1,\tau} - k_{s-1}^P x_{s-1,\tau}) + w_{s\tau}, \qquad s = 1, \ldots, 4, \quad (6.46)$$

where $D_s y_{s\tau} = y_{s\tau} - \alpha_s^y y_{s-1,\tau}$ is the stationary periodic quasi-difference of $y_{s\tau}$ and $w_{s\tau}$ is stationary (but possibly periodic). The adjustment coefficients λ_s^P will generally be periodic. This ECM could be augmented by appropriate lagged stationary variables, namely $\Delta_1 x_{s-i,\tau}$ $(i = 0, 1, \ldots)$ and $D_{s-i} y_{s-i,\tau}$ $(i = 1, 2, \ldots)$. Current $\Delta_1 x_{s\tau}$ is permitted since this variable is assumed exogenous.

The second case is similar, except that the underlying nonstationary process driving the nonstationarity in y_t derives from the quarter-to-quarter $PI(1)$ process in x_t. Thus, stationary $D_{s-i} x_{s-i,\tau}$ $(i = 0, 1, \ldots)$ and $D_{s-i} y_{s-i,\tau}$ $(i = 1, 2, \ldots)$ may also be included, with the periodic integration coefficients α_s being equal for $y_{s\tau}$ and $x_{s\tau}$ in the special case of nonperiodic cointegration $(k_1^P = \cdots = k_4^P)$.

Case 3 is different because there are four separate underlying unit root processes, one associated with each of the four quarters of the year. These imply separate ECMs, so that

$$\Delta_4 y_{s\tau} = \lambda_s^P (y_{s,\tau-1} - k_s^P x_{s,\tau-1}) + w_{s\tau}, \qquad s = 1, \ldots, 4, \quad (6.47)$$

Additional seasonally differenced $\Delta_4 y_{t-i}$ $(i = 1, 2, \ldots)$ and $\Delta_4 x_{t-i}$ $(i = 0, 1, \ldots)$ may be included. This is, in fact, the case that has generally been considered in practice, as in Birchenhall et al. (1989), Franses and Kloek (1995), Herwartz (1997), and, from a theoretical perspective, Boswijk and Franses (1995). None of these papers, however, provides a rationale of the use of the seasonal difference ECM form (6.47) rather than the periodic difference one of (6.46). The analysis here, however, indicates that the univariate properties of the series y_t, x_t might be used to guide this choice.

6.5.2 Vector Approaches to Periodic Cointegration

One approach to the various possibilities for cointegration that arise for seasonal time series processes is to examine unit root properties within the context of the VAR for the stacked annual vector, as proposed by Franses (1995a). With two variables and quarterly data, this vector is $Z_\tau = (y_{1\tau}, x_{1\tau}, y_{2\tau}, x_{2\tau}, y_{3\tau}, x_{3\tau}, y_{4\tau}, x_{4\tau})'$. The simplest situation occurs when the relationships are captured by the VAR(1) process

$$\Delta_4 Z_\tau = A Z_{\tau-1} + U_\tau, \tag{6.48}$$

and the properties of the 8×8 matrix A reflect the cases of interest. Here (6.48) represents a bivariate form of the vector representation used in the multivariate test of Franses, considered in Subsection 6.4.4. In addition to cases associated with periodic cointegration, this VAR process is sufficiently general to incorporate seasonal cointegration discussed in Chapter 3; see Franses (1996b) and Subsection 6.5.3 below.

With (periodic or nonperiodic) cointegration between two processes x_t, y_t each of which is $I(1)$ or $PI(1)$, then the rank of A is 7. This follows since there are three cointegrating relationships between the four $x_{s\tau}$ ($s = 1, \ldots, 4$) and four cointegrating relationships between $y_{s\tau}$, $x_{s\tau}$ (see Subsection 6.4.4). An alternative, and equivalent, view is that there are three cointegrating relationships between $x_{s\tau}$ ($s = 1, \ldots, 4$) and another three between $y_{s\tau}$ ($s = 1, \ldots, 4$), with cointegration between all four $y_{s\tau}$, $x_{s\tau}$ following from any single cointegrating relationship.

For full periodic cointegration between two $SI(1)$ processes, A has rank 4, with cointegrating relationships of the form $y_{s\tau} - k_s^P x_{s\tau}$ with not all k_s^P equal over $s = 1, \ldots, 4$. When the 4×8 matrix of cointegrating vectors is denoted by the rows of K', then

$$K' = \begin{bmatrix} 1 & -k_1^P & 0 & 0 & 0 & 0 & 0 & 0 \\ 0 & 0 & 1 & -k_2^P & 0 & 0 & 0 & 0 \\ 0 & 0 & 0 & 0 & 1 & -k_3^P & 0 & 0 \\ 0 & 0 & 0 & 0 & 0 & 0 & 1 & -k_4^P \end{bmatrix}. \tag{6.49}$$

The special case of $k_1^P = k_2^P = k_3^P = k_4^P$ leads to nonperiodic cointegration between the SI (1) processes. With K established as the matrix of cointegrating vectors, the null hypothesis $k_1^P = k_2^P = k_3^P = k_4^P$ can be tested as in Johansen and Juselius (1990). Rejection of this null hypothesis then implies that cointegration should be considered as periodic rather than nonperiodic. It is also clear that partial periodic (or nonperiodic) cointegration can also be handled in this framework, with the number of seasons where cointegration holds giving the rank of A.

For the cointegrating matrix K in (6.49) above and with the assumption of exogenous $x_{s\tau}$, the corresponding ECM is written in scalar terms in (6.47). This implies the 8×4 adjustment matrix Λ in the VAR, where

$$
\Lambda = \begin{bmatrix}
\lambda_1^P & 0 & 0 & 0 \\
0 & 0 & 0 & 0 \\
0 & \lambda_2^P & 0 & 0 \\
0 & 0 & 0 & 0 \\
0 & 0 & \lambda_3^P & 0 \\
0 & 0 & 0 & 0 \\
0 & 0 & 0 & \lambda_4^P \\
0 & 0 & 0 & 0
\end{bmatrix}
\tag{6.50}
$$

and $A = \Lambda K'$.

Although this type of analysis is of theoretical interest, proceeding by the application of the Johansen (1988) VAR methodology is problematic in practice in the present context. This is simply because of the large number of parameters involved in (6.48), with sixty-four elements in A in even the simplest quarterly case where only two variables are involved. Additional variables to be considered in the system would obviously add to the dimension of A, as would the use of monthly data.

The Johansen approach becomes more tractable if the block diagonality of A, implied by (6.49) and (6.50), is exploited. Specifically, when season s is examined we are then interested in the VAR connecting $y_{s\tau}$ and $x_{s\tau}$. This can be written as

$$
\Delta_4 Z_{s\tau} = A_s Z_{s,\tau-1} + U_{s\tau},
\tag{6.51}
$$

where $Z_{s\tau} = (y_{s\tau}, x_{s\tau})'$, A_s is the sth 2×2 diagonal block of A and $U_{s\tau} = [\varepsilon_{s\tau}^y, \varepsilon_{s\tau}^x]'$ is the 2×1 vector of disturbances for the variables in season s of year τ. Clearly, cointegration between $y_{s\tau}, x_{s\tau}$ implies that A_s has rank 1, and this can be examined by using the methodology of Johansen (1988). However, a test of periodic versus nonperiodic cointegration would require the simultaneous examination of (6.51) for $s = 1, \ldots, 4$, which leads back to the system (6.48).

Rather than a VAR approach, the studies of Birchenhall et al. (1989) and Franses and Kloek (1995) apply the Engle–Granger (1987) two-step methodology, with the first step being a test for cointegration between $y_{s\tau}$ and $x_{s\tau}$ using the residuals from the ordinary least-squares regression of $y_{s\tau}$ on $x_{s\tau}$ for the specific s. If cointegration is found, the second step is then estimation of the adjustment coefficient λ_s^P. Although relatively straightforward, this season by season approach cannot examine periodic versus nonperiodic cointegration.

Boswijk and Franses (1995) take a related approach that is effectively based on the ECM for season s, namely (6.47). Their formulation does, however, allow consideration of issues of periodic versus nonperiodic relationships. We return to this approach in Subsection 6.5.4 below. First, however, we use the

VAR approach to consider a more fundamental question, namely whether cointegration between $y_{s\tau}$ and $x_{s\tau}$ should be analyzed as seasonal cointegration (considered in Chapter 3) or as periodic cointegration.

6.5.3 Nonperiodic, Periodic, and Seasonal Cointegration

Franses (1993, 1995a) discusses the practical question of how to select between the seasonal cointegration and periodic cointegration approaches to modeling the long run relationship between nonstationary seasonal time series. For simplicity, and without losing anything of substance, we continue to assume that only the two variables y_t, x_t are involved. As discussed in Chapter 3, the concept of seasonal cointegration is relevant when both the quarterly variables y_t and x_t are seasonally integrated.

As already discussed, full periodic cointegration requires that the cointegrating matrix takes the form of (6.49). From our earlier analysis of seasonal cointegration (Section 3.6), it follows that seasonal cointegration at the zero and both seasonal frequencies implies that the matrix of cointegrating vectors[5] is

$$
K' = \begin{bmatrix}
1 & -k_1 & 1 & -k_1 & 1 & -k_1 & 1 & -k_1 \\
-1 & k_2 & 1 & -k_2 & -1 & k_2 & 1 & -k_2 \\
0 & k_I & -1 & k_R & 0 & -k_I & 1 & -k_R \\
1 & -k_R & 0 & k_I & -1 & k_R & 0 & -k_I
\end{bmatrix}. \tag{6.52}
$$

The first row here corresponds to cointegration at the zero frequency and the second to cointegration at the semiannual frequency, while the final two rows capture the two cointegrating relationships for the annual frequency. When the notation of Chapter 3 is used, the first of these annual frequency relationships connects $y_{4\tau}^{(3)} = y_{4\tau} - y_{2\tau}$ to $x_{4\tau}^{(3)} = x_{4\tau} - x_{2\tau}$ and $x_{3\tau}^{(3)} = x_{3\tau} - x_{1\tau}$, while the second connects $y_{3\tau}^{(3)} = y_{3\tau} - y_{1\tau}$ to $x_{1\tau}^{(3)}$ and $x_{3\tau}^{(3)}$. It might be anticipated that $x_{2\tau}^{(3)}$ rather than $x_{4\tau}^{(3)}$ would appear in the latter case, but this replacement is valid because $(1 + L^2)x_{4\tau}^{(3)}$ is stationary.

As pointed out by Franses (1993, 1995a), periodic and seasonal cointegration imply different restrictions on the matrix of cointegrating vectors. This can be seen from a comparison of (6.49) and (6.52), since no linear combination

---

[5] Franses (1995a) also writes out the cointegrating vectors for seasonal cointegration in an analogous way, but (allowing for notational differences) his cointegrating vectors corresponding to the annual seasonal frequency are different from ours through the inclusion of coefficients on $y_{3\tau}^{(3)} = y_{3\tau} - y_{1\tau}$ and $y_{4y}^{(3)} = y_{4\tau} - y_{2\tau}$ in the first and second of these, respectively. These arise because Franses uses the approach of Engle et al. (1993) in which an identification problem is present for the coefficients of these cointegrating relationships. There is no identification issue when our approach of Chapter 3 is used.

of the rows of the former yield the rows of the latter matrix provided the equality $k_1^P = k_2^P = k_3^P = k_4^P$ does not hold. Thus, Franses (1995a) proposes that periodic and seasonal cointegration might be formally distinguished by testing the restrictions implied by each of these types of cointegration. Neither form is nested in the other, so the tests should be conducted in the framework of a general matrix of four cointegrating vectors. Indeed, nonperiodic cointegration, which results from $k_1^P = k_2^P = k_3^P = k_4^P$, can also be tested in this framework. In all cases, once the number of cointegrating relationships has been established as four, these tests of restrictions can be anticipated to follow conventional F distributions under the null hypothesis of the validity of the restrictions. Since the periodic and seasonal forms of cointegration are not nested, there seems no reason to presume initially that the cointegration is periodic, as done in the earlier approach of Franses (1993).

It is clear that nonperiodic cointegration is a restricted case of periodic cointegration. Nonperiodic cointegration is, however, also a special case of seasonal cointegration, and the following proposition establishes a useful equivalence:

Proposition 6.1: For the quarterly variables y_t, $x_t \sim SI(1)$, full nonperiodic cointegration is equivalent to seasonal cointegration that applies at the zero and seasonal frequencies with $k_1 = k_2 = k_R$ and $k_I = 0$.

To see this, note first that nonperiodic cointegration implies the cointegrating matrix

$$K' = \begin{bmatrix} 1 & -k & 0 & 0 & 0 & 0 & 0 & 0 \\ 0 & 0 & 1 & -k & 0 & 0 & 0 & 0 \\ 0 & 0 & 0 & 0 & 1 & -k & 0 & 0 \\ 0 & 0 & 0 & 0 & 0 & 0 & 1 & -k \end{bmatrix}. \tag{6.53}$$

Premultiplication of this matrix by the nonsingular matrix

$$R' = \begin{bmatrix} 1 & 1 & 1 & 1 \\ -1 & 1 & -1 & 1 \\ 0 & -1 & 0 & 1 \\ 1 & 0 & -1 & 0 \end{bmatrix}$$

yields the special case of (6.52) with $k_1 = k_2 = k_R = k$ and $k_I = 0$. Since the matrix R' is nonsingular, it is obvious that this special case of seasonal cointegration also implies nonperiodic cointegration, and hence the two forms of cointegration are equivalent. It might be also noted that the matrix R' here has already been discussed at some length in Chapters 2 and 3.

Thus, nonperiodic cointegration provides a link between the concepts of full periodic and seasonal cointegration, since both the latter forms have nonperiodic cointegration as a special case. This link does not seem to have been noted in the previous literature. Indeed, it might be remarked that in considering cointegration for seasonal nonstationary variables, investigators appear to take an a priori position of whether seasonal or periodic cointegration is appropriate for their variables. The discussion here indicates that when the variables are seasonally integrated, this may not be justified. It seems that further research is warranted on the issues surrounding seasonal versus periodic cointegration.

6.5.4   The Boswijk–Franses Test for Periodic Cointegration

Boswijk and Franses (1995) investigate periodic cointegration for the case in which x_t is allowed to be a vector, but we consider only the situation in which it is univariate in order to keep the discussion relatively simple. Although not all aspects of their analysis require an exogeneity assumption for x_t, we will continue to make this assumption for simplicity of presentation.

Starting with given s, we can use the approach of Boswijk (1994) to test the null hypothesis $\lambda_s^P = 0$ in (6.47). The Boswijk test is for cointegration in a nonseasonal context, but it is valid here since we are considering only the series $y_{s\tau}$ and $x_{s\tau}$ for a specific season s. The case of interest under the alternative hypothesis is stationarity, which implies $\lambda_s^P < 0$. We can note that $\lambda_s^P = 0$ implies that both coefficients are zero, that is $\lambda_s^P = \lambda_s^P k_s^P = 0$. The Boswijk test uses a joint F-type statistic for this null hypothesis, so that a one-sided test is not possible. For notational ease, we rewrite the periodic ECM as

$$\Delta_4 y_{s\tau} = \beta_s' Z_{s,\tau-1} + \varepsilon_{s\tau}^y, \tag{6.54}$$

where $\beta_s' = [\lambda_s^P, -\lambda_s^P k_s^P]$, $Z_{s\tau}$ is as defined above, and (again for simplicity) we assume that the disturbance $\varepsilon_{s\tau}^y$ is i.i.d.$(0, \sigma^2)$. For ease of presentation, we assume that σ^2 is constant over $s = 1, \ldots, 4$.

Note that under the null hypothesis, the univariate annually sampled process $y_{s\tau}$ ($\tau = 1, \ldots$) is a random walk. Therefore, the key asymptotic distribution results for random walk processes, as discussed in Chapter 3, continue to apply. Restated for convenience here in a quarterly context and using a superscript to denote the variable of interest, we see that these are

$$T_\tau^{-1} \sum_{\tau=1}^{T_\tau} \varepsilon_{s\tau}^y y_{s,\tau-1} \Rightarrow \sigma^2 \int_0^1 W_s^y(r)\, dW_s^y(r), \tag{6.55}$$

$$T_\tau^{-2} \sum_{\tau=1}^{T_\tau} y_{s,\tau-1}^2 \Rightarrow \sigma^2 \int_0^1 \left[W_s^y(r) \right]^2 dr, \tag{6.56}$$

where $W_s^y(r)$ ($s = 1, \ldots, 4$) are independent standard Brownian motions, each associated with the random walk $y_{s\tau}$ for the corresponding season s. Now, by assumption, $x_t \sim SI(1)$, and in the spirit of keeping the arguments as simple as possible, we will assume that x_t is a seasonal random walk process, so that $x_{s\tau}$ $(\tau = 1, \ldots)$ is also a random walk. If the disturbance process for $x_{s\tau}$ has variance ω^2 (assumed constant over s), then corresponding to (6.56) we have

$$T_\tau^{-2} \sum_{\tau=1}^{T_\tau} x_{s,\tau-1}^2 \Rightarrow \omega^2 \int_0^1 \left[W_s^x(r) \right]^2 dr, \tag{6.57}$$

where $W_s^x(r)$ are independent standard Brownian motions over $s = 1, \ldots, 4$. With $x_{s\tau}$ exogenous, the random walk processes for $y_{s\tau}$ and $x_{s\tau}$ are independent. Therefore, under the null hypothesis, $W_s^y(r)$ and $W_s^x(r)$ are independent, so that

$$T_\tau^{-1} \sum_{\tau=1}^{T_\tau} \varepsilon_{s\tau}^y x_{s,\tau-1} \Rightarrow \sigma\omega \int_0^1 W_s^x(r) \, dW_s^y(r), \tag{6.58}$$

$$T_\tau^{-2} \sum_{\tau=1}^{T_\tau} x_{s,\tau-1} y_{s,\tau-1} \Rightarrow \sigma\omega \int_0^1 W_s^x(r) W_s^y(r) \, dr. \tag{6.59}$$

In this context, we will also use the vector notation $W_s(r) = [W_s^y(r), W_s^x(r)]'$.

When the parameters of the ECM (6.54) are estimated by ordinary least squares, the Wald test statistic for $\beta_s = 0$ is

$$\text{Wald}(\widehat{\beta}_s) = \widehat{\beta}_s'[\widehat{V}(\widehat{\beta}_s)]^{-1}\widehat{\beta}_s, \tag{6.60}$$

where $\widehat{V}(\widehat{\beta}_s) = \widetilde{\sigma}^2[\sum_{\tau=1}^{T_\tau} Z_{s,\tau-1} Z_{s,\tau-1}']^{-1}$ is the covariance matrix estimator for $\widehat{\beta}_s$ and $\widetilde{\sigma}^2$ is the usual OLS estimator of σ^2. Under the null hypothesis $\beta_s = 0$,

$$\text{Wald}(\widehat{\beta}_s) = \widetilde{\sigma}^{-2} \left[\sum_{\tau=1}^{T_\tau} \varepsilon_s^y Z_{s,\tau-1} \right]' \left[\sum_{\tau=1}^{T_\tau} Z_{s,\tau-1} Z_{s,\tau-1}' \right]^{-1} \left[\sum_{\tau=1}^{T_\tau} \varepsilon_s^y Z_{s,\tau-1} \right]$$

$$\Rightarrow \left[\int_0^1 W_s'(r) \, dW_s^y(r) \right] \left[\int_0^1 W_s(r) W_s'(r) \, dr \right]^{-1}$$

$$\times \left[\int_0^1 W_s(r) \, dW_s^y(r) \right]. \tag{6.61}$$

This distributional result can be established by noting that $\widetilde{\sigma}^2 \xrightarrow{p} \sigma^2$, and using (6.55)–(6.59) after scaling of the terms in the first line of (6.61) by T_τ^{-1} or T_τ^{-2} as appropriate. These scaling terms cancel, as do the scalar variances σ^2 and ω^2. The distribution in (6.61) is identical to that of Boswijk (1994).

In addition to the testing of cointegration for individual seasons, the overall null hypothesis of no cointegration may be considered. This overall null hypothesis is parameterized as $\beta_s = 0, s = 1, \ldots, 4$, and it implies that $y_{s\tau}$ and $x_{s\tau}$ are not cointegrated for any quarter $s = 1, \ldots, 4$. It should be noted that the Brownian motions $W_s^y(r)$ and $W_s^x(r)$ are all mutually independent under this null hypothesis. This arises from arguments corresponding to those of Chapter 3 that each (quarterly) seasonal random walk implies four separate random walks, in this case for each of $y_{s\tau}$ and $x_{s\tau}$ over $s = 1, \ldots, 4$, together with the implication of this overall null hypothesis that y_t and x_t are independent. Therefore, the statistics $\mathrm{Wald}(\widehat{\beta}_s)$ are independent over $s = 1, \ldots, 4$ and an overall statistic can be constructed as

$$\mathrm{Wald}(\widehat{\beta}) = \sum_{s=1}^{4} \mathrm{Wald}(\widehat{\beta}_s)$$

$$\Rightarrow \sum_{s=1}^{4} \left[\int_0^1 W_s'(r) dW_s^y(r) \right] \left[\int_0^1 W_s(r) W_s'(r) \, dr \right]^{-1}$$

$$\times \left[\int_0^1 W_s(r) dW_s^y(r) \right]. \tag{6.62}$$

Boswijk and Franses (1995) provide tables of critical values for the $\mathrm{Wald}(\widehat{\beta}_s)$ and $\mathrm{Wald}(\widehat{\beta})$ statistics.

Although Boswijk and Franses obtain the distributional result in (6.62), their route to this is different. Still leaving aside possible augmentation dynamics, then in our notation they consider the equation

$$\Delta_4 y_t = \sum_{s=1}^{4} \delta_{st} \lambda_s^P \left(y_{t-4} - k_s^P x_{t-4} \right) + \varepsilon_t^y, \qquad t = 1, \ldots, T,$$

where δ_{st} is, as previously, a seasonal dummy variable taking the value one when t falls in quarter s and zero otherwise. This is, of course, simply (6.54) aggregated over the four quarters $s = 1, \ldots, 4$. This last equation can also be written as

$$\Delta_4 y_t = \beta' X_{t-4} + \varepsilon_t^y, \qquad t = 1, \ldots, T, \tag{6.63}$$

with $\beta' = (\beta_1', \beta_2', \beta_3', \beta_4')$ and β_s previously defined, together with $X_t' = (\delta_{1t} y_t, \delta_{1t} x_t, \ldots, \delta_{4t} y_t, \delta_{4t} x_t)$. By construction (that is, through the seasonal dummy variables δ_{st}), the cross-product matrix $\sum_{t=1}^{T} X_{t-4} X_{t-4}'$ is block diagonal, with the sth diagonal block being $\sum_{\tau=1}^{T_\tau} Z_{s,\tau-1} Z_{s,\tau-1}'$. Therefore, the ordinary least-squares coefficient estimates resulting from (6.63) are identical to those of (6.54), and the distributional results of (6.61) and (6.62) continue to apply in the context of (6.63).

If cointegration is indicated between $y_{s\tau}$ and $x_{s\tau}$ for some $s = 1, \ldots, 4$, then Boswijk and Franses suggest estimating the ECM coefficients as

$$\widehat{\lambda}_s^P = \widehat{\beta}_{s1}, \qquad \widehat{k}_s^P = -\widehat{\beta}_{s2}/\widehat{\beta}_{s1}, \tag{6.64}$$

where $\widehat{\beta}_s' = (\widehat{\beta}_{s1}, \widehat{\beta}_{s2})$. When cointegration is implied for all s, then periodicity can be examined for either the λ_s^P or the k_s^P, or both. Thus, having established cointegration, the null hypothesis that this cointegration is nonperiodic can be tested against the periodic alternative by computing an F statistic for the null hypothesis $k_1^P = k_2^P = k_3^P = k_4^P$. However, since $\widehat{k_s^P}$ in (6.64) is obtained as a nonlinear function of the estimated coefficients of (6.63), then (if λ_s^P is allowed to be periodic) nonlinear least-squares estimation is required under the null hypothesis. The key result, however, is that the F distribution applies asymptotically, so that no special tables are required for this test. In effect, this is because cointegration is not at issue.

In a similar way, the null hypotheses $\lambda_1^P = \lambda_2^P = \lambda_3^P = \lambda_4^P$ or $\lambda_1^P = \lambda_2^P = \lambda_3^P = \lambda_4^P$ and $k_1^P = k_2^P = k_3^P = k_4^P$ can be tested for (6.63) by using the usual F statistic. In these cases, only linear least-squares estimations are required. The former implies a test of the null hypothesis $\beta_{11} = \beta_{21} = \beta_{31} = \beta_{41}$, while the latter implies a joint test of this restriction together with $\beta_{12} = \beta_{22} = \beta_{32} = \beta_{42}$. Not surprisingly, both of these test statistics follow conventional F distributions.

6.5.5 Generalizations

The models discussed in this section are deficient in a number of respects, whose implications we now briefly discuss.

Firstly, dynamics have been largely ignored beyond the first-order terms included in order to allow cointegration issues to be considered. However, as noted, the inclusion of stationary dynamic terms is asymptotically innocuous. Thus, the inclusion of $\Delta_4 Z_{\tau-i}$ ($i = 1, \ldots,$) in the VAR (6.48) or the inclusion of $\Delta_4 y_{t-i}$ ($i = 1, \ldots$) and/or $\Delta_4 x_{t-i}$ ($i = 0, 1, \ldots$) in the ECM considered for season s would have no effect on the respective asymptotic distributions obtained for the Boswijk–Franses test. Note again that $\Delta_4 x_t$ is permitted because of the exogeneity assumption made. Similarly, the essential results depend on $\Delta_4 x_t$ being $SI(1)$; the special case where this variable is a seasonal random walk has been taken only for ease of exposition.

Secondly, additional variables could be included. As noted, this will increase the dimension of the VAR in (6.48), with consequent practical implications for the dimensions of A. The addition of stationary exogenous variables to the discussion of Subsection 6.5.4 is largely innocuous (at least asymptotically). However, the number of exogenous variables appearing in the cointegrating relationships of that section affects the asymptotic distributions of $\text{Wald}(\widehat{\beta}_s)$ and $\text{Wald}(\widehat{\beta})$, so that the respective critical values depend on this.

Finally, no deterministic terms appear in any of the models where periodic cointegration has been considered, and the asymptotic distributions of Subsection 6.5.4 have implicitly assumed zero starting values for $y_{s\tau}$ and $x_{s\tau}$. As is usual for unit root processes, the inclusion of deterministic components has nontrivial implications for the asymptotic distributions and hence affects the critical values of the cointegration test statistics. As noted by Boswijk and Franses (1995), seasonal intercepts and seasonal trend terms should typically be included.

The former will account for nonzero starting values of the nonstationary processes, while the latter allows unrestricted drift terms for the variables. At the time of writing at least, there does not appear to be any analysis that considers restricting the deterministic terms in the context of periodic cointegration in order to deliver plausible economic implications such as a common long run trend across the seasons.

6.6 Some Comments on Empirical and Monte Carlo Analyses

Periodic autoregressive models have been applied in a number of studies. Based on the theoretical seasonal consumption model of Osborn (1988), the applications of the late 1980s concentrated on consumption [see Osborn et al. (1988), Osborn and Smith (1989)]. Subsequent analyses, including those by Franses and Romijn (1993), Flores and Novales (1997), and Wells (1997), often examine a wider range of series. The general conclusion of these studies is that the behavior of a substantial proportion of macroeconomic time variables is periodic in nature. These studies generally concentrate on modeling the series over the sample period. When forecasts are considered, Wells (1997) and also Novales and de Fruto (1997) find that periodic models often produce less accurate forecasts than nonperiodic models. This stylized result seems to apply even when hypothesis test results point to a periodic process. However, Novales and de Fruto also conclude that important accuracy gains can be made by imposing nonperiodic coefficients across some (but not all) seasons of the periodic model.

The Monte Carlo evidence of Proietti (1998) may throw some light on these issues. He examines the model specification procedure used by Franses and his co-authors, spelled out in Franses and Paap (1994). This procedure first specifies the order of the VAR representation, assuming a periodic process, and later examines whether the process is, indeed, a periodic one. When applied to a simple nonperiodic unobserved components DGP, this procedure tends to select a low-order VAR, and especially a VAR(1) model, with the true null hypothesis of constant coefficients frequently rejected. On the basis of this evidence, Proietti concludes that the empirical evidence in favor of periodic specifications could be spurious.

In line with the early emphasis on consumption for univariate periodic analyses, applications of periodic cointegration have been almost exclusively concerned with the consumption/income relationship [Birchenhall et al. (1989), Franses and Kloek (1995), Franses and Paap (1995b), Herwartz (1997)]. In this context, the study by Herwartz throws some doubt on whether the use of periodic cointegration improves forecast performance in comparison with a nonperiodic specification.

Our analysis of the effects of seasonal adjustment in Chapter 5 did not consider periodic processes and, indeed, there appears to be a dearth of literature on such effects. One exception is Franses (1995b), who uses Monte Carlo techniques to throw some light on this issue. His conclusion is that seasonal adjustment does not remove all the periodic characteristics, and hence he recommends the use of a periodic approach even for seasonally adjusted data.

To conclude this chapter, we may note that periodic processes have some attractive features, and great strides have been made in recent years in establishing an appropriate toolkit for the statistical analysis of such processes. Overall, however, empirical applications in economics are relatively few to date. While evidence of periodicity has been found, it would be foolhardy to conclude at the present time that the majority of important real macroeconomic variables are of this type. It is also true that the empirical analyses have generally been confined to quarterly series. The data requirements of periodic approaches with monthly data (notably those based on a VAR representation) are large. It may be that methods based on restricting periodic processes, perhaps leading to specifications with relatively few additional parameters compared with nonperiodic models, will prove fruitful in the future. In this way, periodic features that are important may be taken into account without the sacrifice of too many degrees of freedom in relation to nonperiodic models.

7 Some Nonlinear Seasonal Models

7.1 Introduction

This chapter is the least comprehensive of all chapters in this book. It has no pretention to represent the current state of the literature. Instead, it is highly selective in its coverage of topics. There are several reasons for the approach taken in this chapter. First, while there have been many developments on the frontier of linear time series analysis of seasonal processes, it is clear that the topic of nonlinear models is very much unsettled at this point. Second, the subject of nonlinear time series models in general does not enjoy the same level of acceptance and agreement as the topic of linear time series models does. Seasonal models are no exception. Moreover, as this book is being written, there is an emerging field of high-frequency financial data that is very much in its early stages of development.

High-frequency data are most commonly, though not exclusively, encountered in finance. These are transaction based, irregularly spaced, and frequently observed time series. They reach the ultimate level of disaggregation, and therefore are sometimes called ultra-high-frequency data [see Engle (2000)]. Seasonality of a different type is a major source of time series fluctuations in high-frequency data. These are not quarterly or monthly seasonals; they are so called intradaily seasonal patterns. The sheer number of data points and the complexity of the seasonal patterns pose still many challenges to time series econometricians. Various aspects of such data, seasonal and nonseasonal, are discussed in Andersen and Bollerslev (1997a, 1997b), Engle and Russell (1997, 1998), Ghysels (2000a), and Ghysels, Gouriéroux and Jasiak (1997, 1998a, 1998b), among others. At this point this area is emerging and unsettled. It is, therefore, not material that is suited to a detailed coverage in a monograph. We will touch, however, on some models that have their roots in the financial econometrics literature.

The next section of this chapter covers a class of models that forms a bridge between the models covered in Chapter 3, namely linear time series unit root processes, and the class of processes known as ARCH type [see, e.g., Bollerslev, Engle, and Nelson (1994) for a survey]. Stochastic unit root processes discussed by McCabe and Tremayne (1995), Granger and Swanson (1997), and Leybourne, McCabe, and Tremayne (1996) bridge the gap. Seasonal stochastic unit root models were analyzed by Taylor and Smith (1998) and have several interesting features, some of them inherited from stochastic unit root models. These processes, which feature a randomized seasonal root with mean unity, display heteroskedastic seasonal integration. Section 7.3 takes us into the world of ARCH-type models. This class of processes aims to capture the clustering of volatility that is so pervasively found in financial time series. We will pay particular attention to the class of seasonal ARCH-type processes that are ARMA-type processes of seasonal heteroskedasticity. It is therefore not surprising that Section 7.4 then covers periodic ARCH-type processes, which share features with the periodic ARMA processes covered in the previous chapter. A final class of models, which also shares properties with periodic ARMA processes, is related to the work of Hamilton (1989) on Markov switching and is treated in the last section.

7.2 Stochastic Seasonal Unit Roots

The processes studied in Chapters 2 and 3 form the bulk of linear seasonal time series analysis. In this section we cover a slight generalization of linear processes, allowing for random parameters. The idea of introducing random coefficients into the framework of linear time series and its relationship with conditional heteroskedasticity has been explored quite extensively by Tsay (1987), who introduced a class of ARMA models with innovations that feature random coefficient autoregressive dynamics. This class of models has links with (1) the ARCH class of processes, which are discussed in the next section, and (2) random coefficient autoregressive models, which are discussed in this section.

Recent work by McCabe and Tremayne (1995), Granger and Swanson (1997) and Leybourne, McCabe, and Tremayne (1996) has introduced autoregressive models with stochastic unit roots. These univariate models are motivated by the idea that not all (macroeconomic) shocks may have the same impact. The initial papers dealing with stochastic unit roots are based on random coefficient processes [Andel (1976)] and are focused exclusively on the nonseasonal case; the seasonal case was examined by Taylor and Smith (1998) and will be covered here.

Consider the following first-order seasonal random coefficient autoregressive process (using the double subscript season/year notation of the previous chapter):

$$(1 - \widetilde{\phi}_{s\tau} L^s)\widetilde{y}_{s\tau} = \varepsilon_{s\tau}, \qquad s = 1, \ldots, S, \ \tau = 1, 2, \ldots, T_\tau, \qquad (7.1)$$

where the coefficient $\widetilde{\phi}_{s\tau} = 1 + \widetilde{\alpha}_{s\tau}$ and $\widetilde{\alpha}_{s\tau}$ is a random process that is a seasonal AR process, namely

$$\widetilde{\alpha}_{s\tau} = \rho \widetilde{\alpha}_{s,\tau-1} + \xi_{s\tau}, \tag{7.2}$$

with $0 \leq \rho \leq 1$ and $\widetilde{\alpha}_{s0} = 0$ for $s = 1, \ldots, S$. Both innovations are i.i.d. and normally distributed, $\varepsilon_{s\tau} \sim N(0, \sigma^2)$ and $\xi_{s\tau} \sim N(0, \omega^2)$. From (7.1) and (7.2) one obtains a randomized seasonal autoregressive process, namely

$$\Delta_S \widetilde{y}_{s\tau} = \widetilde{\alpha}_{s\tau} \widetilde{y}_{s,\tau-1} + \varepsilon_{s\tau}. \tag{7.3}$$

Here (7.3) implies that the seasonally differenced process is conditionally heteroskedastic, namely $\Delta_S \widetilde{y}_{s\tau}$ conditional on its own past is normally distributed $N(\rho \widetilde{\alpha}_{s,\tau-1} \, \widetilde{y}_{s,\tau-1}, \, \sigma^2 + \omega^2 \widetilde{y}_{s,\tau-1}^2)$. Taylor and Smith (1998) therefore call it a heteroskedastic seasonally integrated process. In the degenerate case where $\omega^2 = 0$, the process $\widetilde{y}_{s\tau}$ is a regular seasonal random walk with homoskedastic innovations. This leads to the following hypotheses:

$$H_0 : \omega^2 = 0, \qquad H_A : \omega^2 > 0,$$

for which a locally most powerful test is proposed.

Before discussing the test, we should note that the authors introduce a richer class of processes, in which the observed values $y_{s\tau}$ are related to $\widetilde{y}_{s\tau}$ through

$$y_{s\tau} = \phi(L)[\widetilde{y}_{s\tau} - \mu_s^* - \beta_s^*(S\tau + s)], \tag{7.4}$$

where $\phi(L)$ is a stationary AR(p) lag polynomial and μ_s^* and β_s^* are seasonally varying deterministic components. Under $H_A : \omega^2 > 0$ we can estimate the model in (7.4) by means of maximum likelihood by using the Kalman filter, as discussed in Leybourne, McCabe, and Mills (1996). They also adopt the analysis of Cox (1983) and Chesher (1984) to develop a locally most powerful test for constancy of parameters against a general random coefficient specification. Taylor and Smith (1998) propose a test for $H_0 : \omega^2 = 0$ using the same principle. The Taylor–Smith test for heteroskedastic seasonal integration can be written as

$$S_\rho = \frac{1}{\sigma^4} \sum_{s=1}^{S} \sum_{j=1}^{T_\tau} \rho^{-2j} \left[\left(\sum_{\tau=j}^{T_\tau} \rho^\tau y_{s,\tau-1} \Delta_S y_{s\tau} \right)^2 - \sigma^2 \sum_{\tau=j}^{T_\tau} \rho^{2\tau} y_{s,\tau-1}^2 \right].$$

$$\tag{7.5}$$

The statistic depends on a nuisance parameter ρ that defines the seasonal AR(1) process in (7.2). This parameter is unidentified under the null. A simple solution to remedy the presence of ρ is to compute S_0 and S_1, that is, the two polar cases with $\rho = 0$ and $\rho = 1$. The limiting distributions for the Taylor–Smith S_0 and S_1 statistics are nonstandard. Critical values are provided by the authors for the cases $S = 2, 4$, and 12.

It was noted above that the seasonal heteroskedasticity of $\tilde{y}_{s\tau}$ (and hence of $y_{s\tau}$) is of the form $\sigma^2 + \omega^2 \tilde{y}_{s,\tau-1}$. Hence, under the alternative $\omega^2 > 0$ the process exhibits explosive seasonal heteroskedasticity, since $\tilde{y}$ is not a covariance stationary process. This is a relatively unattractive property of this class of processes. In the next section we consider processes with more attractive seasonal heteroskedasticity features.

7.3 Seasonal ARCH Models

The class of ARCH models, introduced by Engle (1982), has been widely used and extensively studied. Bollerslev, Chou, and Kroner (1992) provided a survey of empirical applications in finance, and important theoretical developments were surveyed by Bera and Higgins (1993) and Bollerslev, Engle, and Nelson (1994). The nonseasonal GARCH(1,1) of Bollerslev (1986) is the benchmark model in financial time series applications. In the first subsection, we will digress on the properties of seasonal GARCH models. In the second subsection, we will reexamine issues related to Chapter 5, namely how seasonal adjustment filters have an impact on the volatility dynamics.

7.3.1 The Class of Models

The case of seasonal GARCH models fits the generic class of ARCH discussed by Bollerslev et al. (1994). Namely, consider a process y_t with conditional mean:

$$m_t(\beta_0) = E_{t-1}(y_t),$$

with parameter vector β, taking true value β_0, and conditional variance

$$\sigma_t^2(\beta_0) = \text{Var}_{t-1}[y_t - m_t(\beta_0)].$$

The parameter β_0 will be suppressed, and without loss of generality we will also set $m_t = 0$ for convenience. The class of GARCH(p, q) models can be written as

$$\sigma_t^2 \equiv \omega + \phi(L)\varepsilon_{t-1}^2 + \theta(L)\sigma_{t-1}^2, \tag{7.6}$$

with $\phi(L)$ of order p and $\theta(L)$ of order q. The GARCH(p, q) process can be rearranged as an ARMA $[\max(p, q), q]$ model for ε_t^2:

$$\varepsilon_t^2 = \omega + [\phi(L) + \theta(L)]\varepsilon_{t-1}^2 - \theta(L)v_{t-1} + v_t, \tag{7.7}$$

where $v_t \equiv \varepsilon_t^2 - \sigma_t^2$. This mapping into ARMA representations yields immediately a connection with seasonal ARMA and ARIMA models.

Before proceeding with the discussion of the properties of seasonal GARCH processes it is worth concentrating first on the empirical applications for which they have been found useful. First and foremost it should be noted that seasonality in financial-market volatility is pervasive. For instance, Gallant, Rossi, and Tauchen (1992) reported that the historical variance of the Standard and Poor's composite stock-price index in October is almost ten times the variance for March. Similar evidence was reported by Schwert (1990) and Glosten, Jagannathan, and Runkle (1993) and was also reported in Section 1 of Chapter 1 (see Figure 1.3). Bollerslev and Hodrick (1995) found evidence for significant seasonal patterns in the conditional heteroskedasticity of monthly stock-market dividend yields. At the daily frequency, French and Roll (1986) and Baillie and Bollerslev (1989) demonstrated that daily stock-return and foreign-exchange-rate volatility tend to be higher following nontrading days, although proportionally less than during the time period of the market closure. At the intraday level, Wood, McInish, and Ord (1985) and Gouriéroux, Jasiak, and LeFol (1999) documented the existence of a distinct U-shaped pattern in the variances of stock returns over the course of the trading day. Equally pronounced patterns in the volatility of intraday foreign-exchange rates were characterized by Baillie and Bollerslev (1991), Harvey and Huang (1991), and Dacorogna et al. (1993).

Although these pronounced daily and intraday seasonal patterns are arguably irrelevant for the analysis of data recorded at lower frequencies, the increased availability of high-frequency financial time series has stimulated a large recent interest in a variety of new modeling issues explicitly related to such data. For instance, several authors have investigated the interrelation between returns in geographically separated financial markets that trade sequentially with little, if any, overlap in their trading hours. The focus of these studies has typically been on the transmission of information as measured by the degree of spillover in the mean returns and/or volatility from one market to the next. Important contributions include those of Engle, Ito, and Lin (1990), who utilized foreign-exchange rates observed at four points during the 24-hour trading day, and Hamao, Masulis, and Ng (1990), who relied on daily opening and closing prices for the stock indexes in three countries.

A second strand of the intraday, time series oriented literature has been concerned with the lead-lag relations between two or more markets that trade simultaneously. Examples include Baillie and Bollerslev (1991), who utilized hourly observations on five-exchange rates, and Chan, Chan, and Karolyi (1991), who investigated 5-minute returns from stock-index and stock-index-futures markets. Finally, a third group of works explored the role of information flow and other microstructure variables as determinants of intraday-return volatility. This literature is exemplified by Bollerslev and Domowitz (1993), who analyzed 5-minute foreign-exchange returns; Locke and Sayers (1993), who modeled 1-minute stock-index-futures returns; Laux and Ng (1993) and Foster and Viswanathan (1995), who relied on half-hourly foreign-exchange and equity returns, respectively; and Goodhart et al. (1993), who investigated quote-by-quote returns from the interbank foreign-exchange market.

Although many new and interesting results have been uncovered in the just-mentioned studies, on closer inspection the conflicting evidence regarding the volatility dynamics obtained across the different sampling frequencies also raises important new questions concerning the most judicious choice of model structure and the proper treatment of the important intraday seasonal volatility patterns; see Andersen and Bollerslev (1997b), Engle and Russell (1998), Ghysels et al. (1998a, 1998b), and Ghysels (2000a), and among others, for further discussion along these lines. Finally, it should also be noted that quite a few macroeconomic (hence not only financial) time series feature seasonality in the variance [see, e.g., Burridge and Wallis (1990), Fiorentini and Maravall (1996), Jaditz (1996) and Racine (1997)].

The simplest seasonal GARCH model is, by analogy to the GARCH(1,1), as follows:

$$\varepsilon_t^2 = \omega + [\theta_1 + \phi_1]\varepsilon_{t-S}^2 - \theta_1 v_{t-S} + v_t, \tag{7.8}$$

corresponding to a seasonal $ARMA_S(1,1)$ process. The process is called integrated when $\theta_1 + \phi_1 = 1$ and resembles the seasonal unit root processes discussed in Chapter 3. It shares more features, however, with integrated GARCH(1,1), or IGARCH(1,1), particularly with regard to strict stationarity. The latter issue will be discussed shortly. GARCH models do not successfully capture thick-tailed returns and volatility clustering, but do exhibit certain other stylized facts encountered in the data. Therefore, many alternative models have been suggested. Several have straightforward seasonal extensions. The EGARCH model of Nelson (1991) was introduced to capture leverage effects. EGARCH models can be written as ARMA processes for $\ell n \, \sigma_t^2$. The ARMA structure can therefore again include seasonal lags. Other cases of ARCH-type models with seasonal variants include the nonlinear ARCH models of Bera and Higgins (1993), the asymmetric ARCH of Engle (1990), the threshold ARCH of Zakoian (1994), and the so-called GJR model of Glosten et al. (1993), who document, as noted earlier, seasonal patterns in broad stock-market indices like the S&P 500.

Estimation and hypothesis testing of GARCH models, including seasonal GARCH, is discussed at length again by Bollerslev et al. (1994) and also at the end of Section 4 of this chapter. One contentious issue is that of unit root nonstationarity. Nelson (1990) shows for the GARCH(1,1) model with polynomials $\phi(L) = \phi_1 L$ and $\theta(L) = \theta_1 L$, see (7.8), that ε_t and σ_t^2 are strictly stationary if and only if $E[\ell n \, (\theta_1 + \phi_1 \varepsilon_t / \sigma_t)] < 0$, a condition that includes IGARCH as strictly stationary. The extension to higher-order GARCH(p, q) processes has been studied by Bougerol and Picard (1991). Their analysis covers seasonal

ARCH models and implies that a seasonal IGARCH(1,1) is strictly stationary under suitable regularity conditions.

7.3.2   The Effects of Filtering on ARCH Models

Many effects produced by seasonal adjustment filters are still not well understood. We have a fairly good grasp of what happens to linear time series models or regression models as discussed in Chapter 5. However, in the case of (1) nonlinear models, (2) nonlinear features of data, or (3) when nonlinear filtering is applied to the data, the implications are to a large extent still unknown and unexplored.

Ghysels, Granger, and Siklos (1998) shed light on the effect of filtering and, in particular, of seasonal adjustment filters on the volatility dynamics of time series. Obviously, standard procedures such as X-11 are not designed to deal with time series exhibiting conditional (seasonal) heteroskedasticity. The effect of filtering raises issues quite similar to those encountered with temporal aggregation. Indeed, linear filtering and temporal aggregation both involve combinations of observations pertaining to different time periods. The class of ARCH processes as introduced by Engle (1982), and generalized by Bollerslev (1986), is not closed under temporal aggregation as noted by Drost and Nijman (1993). For temporal aggregation to work, one has to weaken the original definition of the process. Therefore, we will consider a linear regression model with *weak* GARCH(p, q) disturbances:

$$y_t = x_t'\beta + \varepsilon_t, \tag{7.9}$$

$$\varepsilon_t^2 = \omega + \sum_{j=1}^{\max(p,q)} (\phi_j + \theta_j)\varepsilon_{t-j}^2 + \upsilon_t - \sum_{j=1}^{q} \theta_j \upsilon_{t-j}, \tag{7.10}$$

where a weak GARCH(p, q) process, as defined by Drost and Nijman, implies that the conditional variance σ_t^2 satisfies $\sigma_t^2 = E_{Lt}(\varepsilon_{t+1}^2)$ with $\varepsilon_{t+1}^2 = \sigma_t^2 + \upsilon_{t+1}$, and $E_{Lt}(\cdot)$ is defined as the linear projection on the space spanned by $1, (\varepsilon_{t-j}, \varepsilon_{t-j}^2) : j \geq 0$. Moreover, the process υ_{t+1} is a Martingale difference sequence with respect to the linear span filtration. For later reference we will formalize this, namely:

Assumption 7.3.1: The GARCH process in (7.10) satisfies the conditions of a weak GARCH.

Following Sims (1974) and Wallis (1974), we assume that the regressors x_t are strictly exogenous. Moreover, both y_t and x_t have nonseasonal (*ns*) and seasonal (*s*) components, namely

$$z_t = z_t^{ns} + z_t^s, \qquad z = x, y.$$

It is assumed that all the data (and hence disturbances) are filtered by the same linear filter, that is,

$$\hat{z}_t^{ns} = z_t^F = v(L)z_t = \sum_{k=-\infty}^{+\infty} v_k L^k z_t, \qquad z = x, y,$$

where $\hat{z}^s = z_t - \hat{z}_t^{ns}$, $z = x, y$, and $L^k z_t = z_{t-k}$. Since a uniform filter is used across all data we do not expect any bias in the OLS estimator $\hat{\beta}$ by using y_t^F and x_t^F as Sims and Wallis showed in their seminal papers. To facilitate our presentation we will assume $\hat{\beta} \equiv \beta$ and ignore all estimation uncertainty in order to focus on the properties of ε_t, in particular its volatility dynamics. Hence, we are interested in studying the properties of $\varepsilon_t^F \equiv v(L)\varepsilon_t$, namely the filtered residual process. Let us first consider the autocovariance structure of the squared unfiltered series:

$$\gamma_2(j) = E_L\left(\varepsilon_t^2 \varepsilon_{t-j}^2\right), \tag{7.11}$$

where $E_L(\cdot)$ is the linear (unconditional) projection associated with the L^2 representation of the $\{\varepsilon_t^2\}$ process [see, e.g., Drost and Nijman (1993)]. Its filtered counterpart can be written as

$$\gamma_2^F(j) = E_L\left(\varepsilon_t^F\right)^2\left(\varepsilon_{t-j}^F\right)^2 = E_L(v(L)\varepsilon_t)^2(v(L)\varepsilon_{t-j})^2. \tag{7.12}$$

Below we deal first with the general linear filtering case without being specific about the particular features of the filter weights. In the following parts, we treat some specific special cases.

The General Case of Linear Filters
Using Assumption 7.1 we can rewrite (7.12) and obtain a first general result. In particular, the weak GARCH assumption of Drost and Nijman implies that $E(\varepsilon_t^2 \varepsilon_{t'} \varepsilon_{t''}) = E_L(\varepsilon_t^2 \varepsilon_{t'} \varepsilon_{t''}) = 0$ for $t \neq t' \neq t''$. It will also be convenient to introduce the following notation:

$$v_2(L) \equiv \sum_{k=-\infty}^{+\infty} v_k^2 L^k.$$

Proposition 7.1: Under Assumption 7.1 the autocovariance function of the squares of the filtered process $\varepsilon_t^F \equiv v(L)\varepsilon_t$ satisfies

$$\gamma_2^F(j) = E_L\left(v_2(L)\varepsilon_t^2\right)\left(v_2(L)\varepsilon_{t-j}^2\right)$$
$$+ \sum_{k=-\infty}^{+\infty}\sum_{i<k} v_i v_k v_{i+j} v_{k+j} E_L\left(\varepsilon_{t-i}^2 \varepsilon_{t-j-k}^2\right). \tag{7.13}$$

This proposition is established in Ghysels et al. (1998). It should be noted that the formula in (7.13) is difficult to appraise unless we make some specific assumptions either about the filter weights or the process. The easiest case is one in which there are no GARCH features; that is, ϕ_j and θ_j are both zero.

In this special case of homoskedastic errors, one obtains

$$\gamma_2^F(j) = \left[\sum_{k=-\infty}^{+\infty} v_k^2 v_{k+j}^2 \right] \gamma_2(0), \tag{7.14}$$

which means that filtering a homoskedastic residual process with a general linear filter will yield ARCH-type effects determined by the squared filter weights. Indeed, from (7.14) we can also formulate the autocorrelation function as follows:

$$\rho_2^F(j) = \left[\sum_{k=-\infty}^{+\infty} v_k^2 v_{k+j}^2 \right] \Bigg/ \left[\sum_{k=-\infty}^{+\infty} v_k^4 \right], \tag{7.15}$$

where $\rho_2^F(j)$ is the autocorrelation of the squared filtered residuals. One would like to use some specific values for the filter weights of course. This will be treated in the next subsection. In general, one can say that the autocovariance structure of the squared residuals before and after (linear) filtering resembles somewhat that of linearly filtered ARMA models. The discussion appearing in Section 5.4 could also be applied to transplant those results to the case of weak GARCH(p, q) models, provided one makes two important modifications. The first is that the second term in (7.13) has to be negligible, which is in some cases valid as will be discussed later. Second, unlike the linear ARMA case, we no longer need to investigate the actual filter weights but rather the squared weights. These have never before been the focus of attention of course, even for frequently used filters such as the X-11 filter.

Some Specific Cases of Linear Filters
        The discussion of general linear filters applies to many specific cases such as first and seasonal differencing filters, linear versions of the X-11 filter, to optimal linear signal extraction filters [see, e.g., Pierce (1979), Bell (1984), Maravall (1988), among others] or to filtering procedures often encountered in empirical macroeconomics such as the Hodrick and Prescott (1997) and Baxter and King (1999) high-pass filters. In the remainder of this section, we will focus our attention on some specific filters in order to derive theoretical results that are easier to interpret than (7.13).

Two frequently encountered linear filters will be considered here. The first is the difference filter $v(L) \equiv (1 - L^S)$ where S can take any positive integer value; that is, this class includes first differencing ($S = 1$) as well as seasonal differencing filters ($S > 1$). This covers situations in which the variables are (seasonally) differenced, which would occur when the regression model (7.9) has GARCH disturbances while the variables are nonstationary and the data are filtered before estimating the volatility dynamics. The second case is surely the more common and more interesting. It involves the linear version of the X-11 filter discussed in Chapter 4. The next two propositions cover the filters $v(L) \equiv (1 - L^S)$ and they are stated without proof; see Ghysels, Granger and Siklos (1998) for details.

Proposition 7.2: Let $v(L) \equiv (1 - L^S)$; then under Assumption 7.1 the autocovariance function of the squares of the filtered process satisfies

$$\gamma_2^F(0) = 2\gamma_2(0) + 6\gamma_2(S), \qquad (7.16)$$

while for $j > 0$,

$$\gamma_2^F(j) = 2\gamma_2(j) + \gamma_2(j + S) + \gamma_2(j - S). \qquad (7.17)$$

A fairly simple case of interest is again homoskedastic disturbances, that is, the unfiltered process features $\gamma_2(j) = 0$, for $j > 0$. In such a case $\gamma_2^F(S) = \gamma_2(0) \neq 0$, and we can write the autocorrelation function as $\rho_2^F(j) = 1$ for $j = 0, S$ and zero otherwise. Hence, first differencing introduces ARCH(1) effects in homoskedastic residuals while seasonal differencing produces seasonal ARCH. We can carry this result further, as shown in the following proposition.

Proposition 7.3: Let $v(L) \equiv (1 - L^S)$; then under Assumption 7.1 the autocorrelation function of the squares of the filtered process is unbiased; that is, $\rho_2^F(j) = \rho_2(j)$, if and only if

$$6\rho_2(S)\rho_2(j) - \rho_2(j + S) - \rho_2(j - S) = 0.$$

Next, we turn to the linear X-11 filter. The filter is two sided and involves over eighty leads and lags, which makes the derivation of explicit analytical results, such as in Propositions 7.2 and 7.3, much more difficult. Fortunately we can investigate the features of the linear X-11 filter by other means. First we should note that the second term in (7.13) becomes negligible as the product of adjacent X-11 filter weights is small, a useful feature of the filter. A consequence of this feature, indeed, is that we can simply focus on the features of the $v_2(L)$ filter, that is, a filter with the squared weights of the linear X-11 filter. Since we focus on the effect of filtering, it is worthwhile to consider first a spectral-domain approach.[1] In Figure 7.1, we plot the squared gain (or transfer function) implied by the squared weights of the monthly linear X-11 filter. For the purpose

---

[1] Since we restrict ourselves to linear projections, we can rely on the equivalent time-domain and spectral-domain representations provided that the innovation process has finite variance. In many empirical applications one assumes that v_t has a Student distribution. The estimated degrees of freedom are always larger than two, which justifies our assumption about the innovation variance.

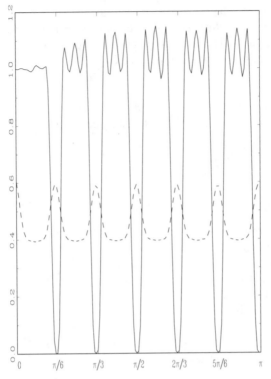

Figure 7.1. Transfer function of the linear monthly X-11 filter and squared weight filter.

of comparison, we also plot the linear X-11 filter transfer functions. Hence, Figure 7.1 shows the transfer functions of $v(L)$ and $v_2(L)$ in the case of the linear X-11 monthly filter.

The transfer functions appearing in Figure 7.1 are quite revealing. The X-11 filter has the familiar pattern that retains the spectral power at all but the seasonal frequency and its harmonies. The filter with squared weights has very different properties. First, as we expect from a smoothing filter, we observe the variance reduction effect. Indeed, the transfer function takes values between roughly 0.4 and 0.62. Another feature to note is that the filter weights of the linear X-11 filter sum to one [a feature important for leaving constants and linear trends unaffected, as stressed by Ghysels and Perron (1993) in the context of unit root testing]. The sum of the squared weights is less than one, more specifically 0.7852, which yields a zero-frequency squared gain of $(0.7852)^2$ or 0.6165, which is the value appearing at the zero frequency in Figure 7.1. The most remarkable feature, however, is that $v_2(L)$ does not have troughs at the seasonal frequency and its harmonies. Instead, it actually has small peaks. Consequently, the X-11 filter, while reducing the overall variance of ε_t, will in fact slightly amplify instead of removing seasonal correlation in the conditional variance dynamics.

The Case of Weak GARCH(1,1)

In the preceding paragraphs we devoted our attention to some specific filters without explicit assumptions regarding the unfiltered volatility structure. We will now assume a specific autocorrelation structure and investigate the effect of linear filtering. The simplest case one can consider is the weak GARCH(1,1) model. Simplicity is not the only reason to elaborate on this particular case. Indeed, it is also appealing to digress on GARCH(1,1) processes because they appear as quite relevant in many empirical applications. Obviously, while the GARCH(1,1) model will figure prominently in our analysis, we also need to examine, given the context of seasonality, GARCH models that exhibit seasonal features. This will be treated in the next section. From Bollerslev (1986) we know that we can rely on standard results of ARMA models [see, e.g., Box and Jenkins (1976) or Fuller (1996)]. Hence, for the weak GARCH(1,1) process we have that

$$\gamma_2(0) = \frac{[1 + \theta^2 - 2(\phi + \theta)\theta]}{[1 - (\phi + \theta)^2]}\sigma_v^2 \tag{7.18}$$

for lag zero and for $j \neq 0$,

$$\gamma_2(j) = \frac{\phi[1 - (\phi + \theta)\theta]}{1 - (\phi + \theta)^2}(\phi + \theta)^{j-1}\sigma_v^2, \tag{7.19}$$

where the subscript has been dropped from ϕ and θ for notational simplicity. The autocorrelation function is therefore

$$\rho_2(j) = \frac{\phi(1 - \phi\theta - \theta^2)}{1 - 2\phi\theta - \theta^2}(\phi + \theta)^{j-1}. \tag{7.20}$$

We can take advantage of these specific autocovariances to obtain more explicit formulas that describe the effect of certain linear filters on the volatility dynamics. Proposition 7.4 states such a result, namely:

Proposition 7.4: Let $v(L) \equiv (1 - L^S)$. Moreover, let us denote $\lambda \equiv (\phi + \theta)$; then under Assumption 7.1 the autocorrelation function of the squares of a filtered GARCH(1,1) process satisfies

$$\rho_2^F(j) \equiv \frac{2 + \lambda + \lambda^S}{2 + 6\lambda\rho_2(S)}\rho_2(j), \tag{7.21}$$

and the autocorrelation function is unbiased if the parameters ϕ and θ solve the following equation:

$$2\lambda^S\theta - (1+\theta^2 - 6\phi\theta)\lambda^{S-1} + 6\phi\lambda^{S-2} + 2\lambda\theta - (1+\theta^2) = 0.$$

$$(7.22)$$

The proof follows straightforwardly from Propositions 7.1 and 7.2. It is interesting to note there will be a downward bias in the autocorrelation function after filtering if (7.22) holds with $<$ instead of equality. Similarly, replacing the equality in (7.22) by $>$ will describe parameter settings for the GARCH(1,1) that feature an upward bias in the autocorrelation function induced by filtering.

Three special cases of (7.22) are most relevant in practical applications. They are the cases $S = 1$, $S = 4$, and $S = 12$. For these cases, the equation specializes (assuming $\lambda \neq 0$) to the following:

$$f_1(\phi, \theta) \equiv 4\lambda^2\theta - 2(1+\theta^2 - 3\phi\theta)\lambda + 6\phi = 0, \qquad (7.23)$$

$$\begin{aligned} f_4(\phi, \theta) \equiv{}& 2\lambda^4\theta - (1+\theta^2 - 6\phi\theta)\lambda^3 + 6\phi\lambda^2 \\ & + 2\lambda\theta - (1+\theta^2) = 0, \end{aligned} \qquad (7.24)$$

$$\begin{aligned} f_{12}(\phi, \theta) \equiv{}& 2\lambda^{12}\theta - (1+\theta^2 - 6\phi\theta)\lambda^{11} + 6\phi\lambda^{10} \\ & + 2\lambda\theta - (1+\theta^2) = 0. \end{aligned} \qquad (7.25)$$

Ghysels et al. (1998) rely on numerical computations to characterize these three equations. Different areas of the parameter space entail different peaks; some are positive, while others are negative. The empirically most relevant case of $\phi + \theta \approx 0.1$ and $\phi \approx 0.1$ shows upward biases. Ghysels et al. (1998) also investigate the actual X-11 program with all its potential sources of nonlinearities and compare it with the linear filter results. Moreover, they consider in addition to GARCH(1,1) also seasonal GARCH processes. Finally, they also compare finite sample properties with the asymptotic ones. For this last case they no longer rely on analytic methods but (have to) rely on Monte Carlo simulations. Focusing on the autocorrelation function of the squared residuals from various GARCH processes, they find that X-11 tends to reduce, and in some cases, completely eliminates seasonal volatility dynamics and substantially reduces the overall persistence. They also found substantial differences between the linear X-11 and actual X-11 filter, showing the significant impact outlier corrections have in practice.

7.4   Periodic GARCH Models

The GARCH(p, q) model, discussed in the previous section, often provides a parsimonious representation of the volatility dynamics in financial time series. We noted that the GARCH(p, q) process may be interpreted as an ARMA

$(\max\{p, q\}, q)$ model for $\{\varepsilon_t^2\}$; see (7.7). Instead of having a fixed parameter structure for the conditional-variance equation, it is possible to draw on the similarity of the ARMA(p, q) model and periodic ARMA processes, and consider a time-varying coefficient model for conditional heteroskedasticity.

To define such structures, consider a modified Borel σ-field filtration in which the usual Ω_{t-1} is augmented by a process defining the stage of the periodic cycle at each point in time, say Ω_{t-1}^s. The class of periodic GARCH, or P-GARCH processes[2], may now be defined as

$$E\left[\widetilde{\varepsilon}_t \middle| \Omega_{t-1}^s\right] = 0, \tag{7.26}$$

$$E\left[\widetilde{\varepsilon}_t^2 \middle| \Omega_{t-1}^s\right] \equiv \widetilde{\sigma}_t^2 = \omega_{s(t)} + \sum_{i=1}^{p} \phi_{is(t)}\widetilde{\varepsilon}_{t-i}^2 + \sum_{j=1}^{q} \theta_{js(t)}\widetilde{\sigma}_{t-j}^2, \tag{7.27}$$

where $s(t)$ refers to the stage of the periodic cycle at time t. The analogy of this equation with the periodic processes examined in Chapter 6 should be apparent. Note that Ω_{t-1}^s appears both in the conditional mean and variance equations. Consequently, $\widetilde{\varepsilon}_t$ may differ from ε_t defined in (7.9). For instance, ε_t may be the residual from a fixed parameter seasonal ARMA model, whereas $\widetilde{\varepsilon}_t$ refers to the residuals from a periodic ARMA model for the conditional mean. Although the lag lengths p and q do not depend on $s(t)$, this entails no loss in generality because p and q may be set to their maximal orders across all stages of the periodic cycle. Of course, for the P-GARCH(p, q) model to be well defined, the conditional variance, $\widetilde{\sigma}_t^2$, must be positive almost surely. Necessary and sufficient conditionals on the $\omega_{s(t)}$, $\phi_{is(t)}$, and $\theta_{js(t)}$ parameters for this to hold true in general are elusive but may be easily verified on a case-by-case basis following the approach of Nelson and Cao (1992).

The most straightforward P-GARCH model is obtained when the periodic cycle is purely repetitive; that is, $s(t) = 1 + [(t - 1) \bmod S]$, where S is the length of the cycle. An example of such a repetitive cycle would be the intraday pattern in market activity associated with the regular opening and closure of financial markets. In many empirical applications to financial time series, however, $s(t)$ may be governed by a variable predetermined cycle with an upperbound S. For example, with daily data, nontrading days usually occur every fifth observation, but some weeks have holidays that interrupt this regular weekly pattern. In this situation $S = 5$, but not all return cycles actually attain 5 consecutive trading days.

In the existing ARCH literature, the modeling of nontrading-day effects has typically been limited to $\omega_{s(t)}$. The representation in (7.27), however, allows

---

[2] Note that the order of subscripts in $\phi_{is(t)}$ and $\theta_{js(t)}$ is the opposite to that used in Chapter 6, where the first subscript indicated the season and the second the lag. The order adopted in the present chapter is a little neater when the explicit dependence $s(t)$ is recognized.

for a much richer dynamic structure in that the innovations that occur over different time periods may have their own distinct impact on the volatility process through the $\phi_{is(t)}$ and $\theta_{js(t)}$ coefficients. In further interpreting these coefficients, the $\phi_{is(t)}$ may be viewed as a measure of the immediate, or direct, impact of any new arrivals, but the smooth long-term evolution in the volatility process is captured by the $\theta_{js(t)}$ coefficients. In many practical applications the periodic variation is therefore naturally constrained to the $\phi_{is(t)}$ coefficients, keeping $\theta_{js(t)} \equiv \theta_j$ constant across all stages of the cycle. Then (7.27) implies that this results in a P-GARCH process with periodicity in the autoregressive part of the model only.

Analogous to the ARMA representation for the GARCH(p, q) model as in (7.7), it is possible to interpret the P-GARCH(p, q) model defined in (7.27) as a periodic ARMA process for $\{\widetilde{\varepsilon}_t^2\}$ with a time-varying but periodic correlation structure,

$$\widetilde{\varepsilon}_t^2 = \omega_{s(t)} + \sum_{i=1}^{\max(p,q)} (\phi_{is(t)} + \theta_{is(t)})\widetilde{\varepsilon}_{t-i}^2 - \sum_{j=-1}^{q} \theta_{js(t)}\widetilde{\upsilon}_{t-j} + \widetilde{\upsilon}_t,$$

(7.28)

where $\widetilde{\upsilon}_t \equiv \widetilde{\varepsilon}_t^2 - E[\widetilde{\varepsilon}_t^2|\Omega_{t-1}^s] \equiv \widetilde{\varepsilon}_t^2 - \widetilde{\sigma}_t^2$. For periodic ARMA models with a purely repetitive cycle of length S, various results from Chapter 6 then become relevant. In particular, based on the studies of Tiao and Grupe (1980) and Osborn (1991), Section 3 of that chapter discussed the nature of the mapping between the periodic ARMA process and a time-invariant seasonal ARMA model with a cycle of length S; (7.28) suggests the existence of a similar mapping for the P-GARCH class of models with a fixed cycle S. The similarities between periodic ARMA and periodic GARCH processes do not carry through straightforwardly, however. The reason is that the GARCH class of processes defined in terms of conditional expectations is not closed under temporal and/or cross-sectional aggregation; see Drost and Nijman (1993) and Nijman and Sentana (1996). This issue also arose in the discussion of Section 3 of the present chapter and, as there, we circumvent these difficulties by considering the wider class of weak GARCH processes. Thus, we generalize Assumption 7.1 slightly so that $\{\widetilde{\varepsilon}_t\}$ is defined to follow a weak P-GARCH process when $\widetilde{\sigma}_t^2$ in (7.27) corresponds to the best linear projection of $\widetilde{\varepsilon}_t^2$ on Ψ_{t-1}^s, the space spanned by $\{1, \widetilde{\varepsilon}_{t-1}, \widetilde{\varepsilon}_{t-2}, \ldots, \widetilde{\varepsilon}_{t-1}^2, \widetilde{\varepsilon}_{t-2}^2, \ldots\}$ augmented by $s(t)$. That is,

$$E[\widetilde{\varepsilon}_t^2 - \widetilde{\sigma}_t^2|\Psi_{t-1}^s] = E[(\widetilde{\varepsilon}_t^2 - \widetilde{\sigma}_t^2)\widetilde{\varepsilon}_{t-i} \mid \Psi_{t-1}^s],$$
$$= E[(\widetilde{\varepsilon}_t^2 - \widetilde{\sigma}_t^2)\widetilde{\varepsilon}_{t-i}^2 \mid \Psi_{t-1}^s] = 0, \qquad i = 1, 2, \ldots.$$

(7.29)

With this assumption, the periodic ARMA representation for the weak P-GARCH model takes the form

$$\widetilde{\varepsilon}_t^2 = \omega_{s(t)} + \left[\phi_{s(t)}(L) + \theta_{s(t)}(L)\right]\widetilde{\varepsilon}_t^2 + \left[1 - \theta_{s(t)}(L)\right]\widetilde{\eta}_t, \qquad (7.30)$$

with $\phi_{s(t)}(L) \equiv \Sigma_{i=1}^{p}\phi_{is(t)}L^i$, $\theta_{s(t)}(L) \equiv \Sigma_{j=1}^{q}\theta_{js(t)}L^j$, and $\widetilde{\eta}_t \equiv \widetilde{\varepsilon}_t^2 - P(\widetilde{\varepsilon}_t^2|\Psi_{t-1}^s)$, where $P(\cdot|\Psi_{t-1}^s)$ denotes the corresponding linear projection. Note that the projections in (7.29) still involve seasonal conditioning so that the autocorrelation structure remains periodic. The specification of a weak P-GARCH model obviously entails an informational loss when compared with the P-GARCH formulation in (7.27), where $\widetilde{\sigma}_t^2$ is defined as the conditional expectation of $\widetilde{\varepsilon}_t^2$ based on the full information set implied by the Borel σ-field filtration Ω_{t-1}^s. By considering the wider class of weak GARCH processes defined previously, however, we find it possible to carry out the mechanics of the Tiao and Grupe (1980) formula. As discussed in Chapter 6, this essentially amounts to the removal of the seasonal conditioning in (7.29) by averaging out the autocorrelation structure across all seasons, thus resulting here in a time-invariant seasonal weak GARCH process. Further details of this averaging procedure, along with an illustrative example, is provided in Bollerslev and Ghysels (1996).

A variety of estimation and testing procedures have been suggested for conducting inference in ARCH-type models; for a more extensive discussion of these procedures, we refer to the survey articles by Bera and Higgins (1993) and Bollerslev et al. (1994). The scope of this section is not to contribute to the basic theory of estimation and hypothesis testing in ARCH models as such. Instead, our aim is merely to comment on some of the specific issues that arise in the estimation and testing of periodic ARCH structures.

To illustrate, let γ denote the vector of unknown parameters for all S seasons; that is, $\gamma \equiv (\gamma_1', \gamma_2', \ldots, \gamma_S')'$, where for the P-GARCH (p, q) model in (7.27), $\gamma_s \equiv (\omega_s, \phi_{1s}, \ldots, \phi_{qs}, \theta_{1s}, \ldots, \theta_{ps})'$ for $s = 1, \ldots, S$. The conditional log-likelihood function for a strong P-GARCH model may then be conveniently written as the sum of the corresponding conditional log likelihoods for each of the S seasonal cycles,

$$L_T\left(\gamma|\Omega_0^s\right) = \sum_{t=1}^{T} l_t\left(\gamma_{s(t)}\right). \qquad (7.31)$$

In particular, assuming the one-step-ahead prediction errors to be conditionally normally distributed, we have

$$l_t\left(\gamma_{s(t)}\right) = -0.5\left\{\ln(2\pi) + \ln\left[\widetilde{\sigma}_t\left(\gamma_{s(t)}\right)\right] + \left[\widetilde{\varepsilon}_t\left(\gamma_{s(t)}\right)\right]^2\left[\widetilde{\sigma}_t^2\left(\gamma_{s(t)}\right)\right]^{-1}\right\}. \qquad (7.32)$$

Under appropriate regularity conditions the maximum likelihood estimator (MLE) for the true parameter vector γ_0, say $\widehat{\gamma}_T$, is obtained by maximizing (7.31) and (7.32). This satisfies

$$T^{1/2}(\widehat{\gamma}_T - \gamma_0) \to N[0, A^{-1}(\gamma_0)], \tag{7.33}$$

where $A(\gamma_0)$ denotes the Hessian evaluated at γ_0. In many applications with high-frequency financial data, the assumption of conditional normality underlying (7.32) may be violated, however. Fortunately, $\widehat{\gamma}_T$ remains consistent under quite general conditions and may be given a quasi-MLE (QMLE) interpretation. Because the outer product of the gradients and the inverse of the Hessian do not cancel out in this situation, the asymptotic covariance matrix for the QMLE takes the form $A(\gamma_0)^{-1} B(\gamma_0) A(\gamma_0)^{-1}$, where $B(\gamma_0)$ denotes the outer product of the gradients; for further discussion along these lines and a formal proof for the GARCH(1,1) model, see Weiss (1986), Bollerslev and Wooldridge (1992), Lumsdaine (1996), and Lee and Hansen (1994). Moreover, the Monte Carlo evidence of Bollerslev and Wooldridge (1992) and Baillie, Bollerslev, and Mikkelsen (1996) indicates that for moderately large sample sizes the accuracy of the QMLE-based inference procedures with conditionally leptokurtic errors is comparable to that of exact MLE. Even though no formal analytical results are available for the weak GARCH case, the simulations of Drost and Nijman (1993) suggest that for large sample sizes, the QMLE procedure is generally very reliable for the estimation of weak GARCH models also.

Bollerslev and Ghysels (1996) illustrate the practical relevance of the P-GARCH model by using daily data on the Deutschemark/U.S. dollar and Deutschemark/pound sterling exchange rates.

7.5   Periodic Markov Switching Models

We conclude this chapter with a discussion of Markov switching models. There are many applications of (nonperiodic) switching regime models, such as Hamilton (1988, 1989, 1990, 1994b), Phillips (1991), Cecchetti and Lam (1992), Albert and Chib (1993), Diebold, Lee, and Weinbach, (1994), Durland and McCurdy (1994), Filardo (1994), Kim (1994), McCulloch and Tsay (1994), and Sichel (1994), among others. Here we focus directly on the so-called periodic Markovian switching regime structure presented by Ghysels (1991, 1994b, 1997b), which was used to investigate the non-uniformity of the distribution of the National Bureau of Economic Research (NBER) business cycle turning points. The discussion will focus first on a simplified illustrative example to present some of the key features and elements of interest.

Consider a univariate time series process denoted $\{y_t\}$. It will typically represent a growth rate of, say, GNP. Furthermore, let $\{y_t\}$ be generated by the following stochastic structure:

$$\{y_t - \mu[(i_t, \mathbf{s}_t)]\} = \phi\{y_{t-1} - \mu[(i_{t-1}, \mathbf{s}_{t-1})]\} + \varepsilon_t, \tag{7.34}$$

with

$$\mu[(i_{t,st})] = \alpha_0 + \alpha_1 i_t + \sum_{s=1}^{S-1} \delta_{st}\alpha_s, \tag{7.35}$$

where $|\phi| < 1$, ε_t is i.i.d. $N(0, \sigma^2)$ and $\mu[\cdot]$ represents an intercept shift function that includes seasonal dummies δ_{st}. If $\mu[(i_t, \mathbf{s}_t)] = \alpha_0 + \sum_{s=1}^{S-1} \delta_{st}\alpha_s$ with $\mathbf{s}_t = t \bmod(S)$, where S is the frequency of sampling throughout the year, for example, $S = 4$ for quarterly sampling, then (7.34) would simply be a standard linear stationary Gaussian AR(1) model with seasonal mean shifts α_s for $s = 1, \ldots, S$. Instead, we assume that the intercept changes according to a Markovian switching regime model, following the work of Hamilton (1989). The "state of the world" process is different, however, from that originally considered by Hamilton. The state of the world is described by $(i_t, \mathbf{s}_t)$, which is a binary stochastic switching regime process $\{i_t\}$ and the seasonal indicator process. We allow the $\{i_t\}$ and $\{\mathbf{s}_t\}$ processes to interact in the following way, assuming that $i_t \in \{0, 1\} \forall t$:

$$\begin{array}{ccc} & 0 & 1 \\ 0 & q(\mathbf{s}_t) & 1 - q(\mathbf{s}_t) \\ 1 & 1 - p(\mathbf{s}_t) & p(\mathbf{s}_t) \end{array} \tag{7.36}$$

where the transition probabilies $q(\cdot)$ and $p(\cdot)$ are allowed to change with $\mathbf{s}_t$, that is, the season. As $\mathbf{s}_t$ is a mod S series, there are of course at most S values for $q(\cdot)$ and $p(\cdot)$; that is, $q(\mathbf{s}_t) \in \{q^1, \ldots, q^s\}$, and $p(\mathbf{s}_t) \in \{p^1, \ldots, p^s\}$, where $q(\mathbf{s}_t) = q^s$ and $p(\mathbf{s}_t) = p^s$ for $\mathbf{s} = \mathbf{s}_t$. Naturally, when

$$p(\cdot) = \overline{p} \text{ and } q(\cdot) = \overline{q}, \tag{7.37}$$

then we obtain the standard homogeneous Markov chain model considered by Hamilton.

The structure presented so far is relatively simple, yet as we shall see, some interesting dynamics and subtle interdependencies emerge. It is worth comparing the AR(1) model with a periodic Markovian stochastic switching regime structure, as represented by (7.34) through (7.37), and the more conventional linear ARMA processes as well as the periodic ARMA models discussed in Chapter 6. The process $\{y_t\}$ is covariance stationary under suitable regularity conditions discussed later. Consequently, it has a linear Wold MA representation. Yet, the time series model presented in (7.34) through (7.37) provides a relatively parsimonious structure that determines nonlinearly predictable MA innovations. In fact, there are two layers beneath the Wold MA representation.

One layer relates to *hidden periodicities,* as described in Tiao and Grupe (1980) and the previous chapter. Typically, such hidden periodicities can be uncovered by means of augmentation of the state space with the augmented system having a linear representation. However, the periodic switching regime model imposes *further structure* even after the hidden periodicites are uncovered. Indeed, there is a second layer that makes the innovations of the augmented system nonlinearly predictable. Hence, the model has nonlinearly predictable innovations and features of hidden periodicities combined.

To develop this more explicitly, let us first note that the switching regime process $\{i_t\}$ admits the following AR(1) representation:

$$i_t = [1 - q(\mathbf{s}_t)] + \lambda(\mathbf{s}_t)i_{t-1} + v_t(\mathbf{s}_t), \tag{7.38}$$

where $\lambda(\cdot) \in \{\lambda^1, \ldots, \lambda^S\}$ with $\lambda(\mathbf{s}_t) \equiv -1 + p(\mathbf{s}_t) + q(\mathbf{s}_t) = \lambda^s$ for $\mathbf{s}_t = \mathbf{s}$. Moreover, conditional on $i_{t-1} = 1$,

$$v_t(\mathbf{s}_t) = \begin{cases} [1 - p(\mathbf{s}_t)] & \text{with probability } p(\mathbf{s}_t) \\ -p(\mathbf{s}_t) & \text{with probability } 1 - p(\mathbf{s}_t) \end{cases}. \tag{7.39}$$

Conditional on $i_{t-1} = 0$,

$$v_t(\mathbf{s}_t) = \begin{cases} -[(1 - q(\mathbf{s}_t))] & \text{with probability } q(\mathbf{s}_t) \\ q(\mathbf{s}_t) & \text{with probability } 1 - q(\mathbf{s}_t) \end{cases}. \tag{7.40}$$

Here (7.38) is a periodic AR(1) model in which all the parameters, including those governing the error process, may take on different values every season. Of course, this is a different way of saying that the state of the world is not only described by $\{i_t\}$ but also $\{\mathbf{s}_t\}$. While (7.38) resembles the periodic ARMA models that were discussed in Chapter 6, it is also fundamentally different in many respects. The most obvious difference is the innovation process, which has a discrete distribution. There are more subtle differences as well, but we shall highlight those as we further develop the model.

Despite the differences, there are many features that (7.38) and the more standard periodic linear ARMA models have in common. Following Gladyshev (1961), we can consider time-invariant representations of (7.38) that are built on stacked, skip-sampled vectors of observations. Let us define the stacked vector of seasons that is sampled at an annual frequency by using our double index notation:

$$\underline{i}_\tau \equiv (i_{1\tau}, i_{2\tau}, \ldots \ldots, i_{S\tau})', \tag{7.41}$$

$$\underline{v}_\tau \equiv (v_{1\tau}, v_{2\tau}, \ldots \ldots, v_{S\tau})'. \tag{7.42}$$

Following (7.38), we can write the DGP for the vector defined in (7.41) as follows:

$$
\begin{bmatrix}
1 & 0 & \cdots & & 0 \\
-\lambda^2 & 1 & \cdots & & 0 \\
\vdots & \ddots & & & \vdots \\
\vdots & & & & \\
0 & & & 1 & 0 \\
0 & \cdots & \cdots & -\lambda^S & 1
\end{bmatrix}
\underline{i}_\tau =
\begin{bmatrix}
1 - q^1 \\
1 - q^2 \\
\vdots \\
\vdots \\
1 - q^S
\end{bmatrix}
+
\begin{bmatrix}
0 & \cdots & \lambda^1 \\
0 & & 0 \\
\vdots & & \vdots \\
\vdots & & \\
0 & \cdots & 0
\end{bmatrix}
\underline{i}_{\tau-1} + \underline{v}_\tau.
$$

$$(7.43)$$

We would like to highlight two features of (7.43) on which we will digress further. The first is the appearance of seasonal mean shifts, that is, what is typically called "deterministic seasonality"; the second is the basis of a time-invariant Wold MA representation for the (scalar) $\{i_t\}$ process described by (7.38). We shall focus first on the latter, followed by a discussion of the former.

Through (7.43) one can derive the Wold representation of $\{i_t\}$ by using the Tiao–Grupe formula. Note that the resulting process will certainly not be represented by an AR(1) process since part of the state space is "missing." The periodic nature of autoregressive coefficients pushes the seasonality into annual lags of the AR polynomial and substantially complicates the MA component. The arguments are entirely analogous to those of Subsection 6.3.3.

Ultimately, we are of course interested in the time series properties of $\{y_t\}$ as it is generated by (7.34) through (7.36) and how its properties relate to linear ARMA and periodic ARMA representations of the same process. Since

$$
y_t = \alpha_0 + \alpha_1 i_t + \sum_{s=1}^{S-1} \delta_{st}\alpha_s + (1 - \phi L)^{-1}\varepsilon_t, \tag{7.44}
$$

and ε_t was assumed Gaussian and independent, we can simply view $\{y_t\} - \sum_{s=1}^{S-1} \delta_{st}\alpha_s$ as the sum of two independent unobserved processes: namely, $\{i_t\}$ and the process $(1 - \phi L)^{-1}\varepsilon_t$. Clearly, all the features just described about the $\{i_t\}$ process will be translated into similar features inherited by the observed process y_t, while the linear time series representation of $y_t - \sum_{s=1}^{S-1} \delta_{st}\alpha_s$ can be obtained from

$$
s_y(z) = \alpha_1^2 s_i(z) + [1/(1 - \phi z)(1 - \phi z^{-1})](\sigma^2/2\pi). \tag{7.45}
$$

This linear representation has hidden periodic properties and inherits also the nonlinear predictable features of $\{i_t\}$.

Let us briefly return to (7.43). We observe that the linear representation has seasonal means shifts that would appear as a "deterministic seasonal" in the univariate representation of y_t. Hence, besides the spectral density properties appearing in (7.45), which may or may not show peaks at the seasonal frequency, we note that periodic Markov switching produces seasonal mean shifts in the univariate representation. This result is, of course, quite interesting since intrinsically we do have a purely random stochastic process with occasional mean shifts. The fact that we obtain something that resembles a deterministic seasonal component simply comes from the unequal propensity to switch regime (and hence mean) during some seasons of the year. It should also be noted of course that seasonal means shifts appear in (7.35) already. Consequently, these mean shift coefficients α_s in (7.35) are expected to differ from seasonal dummies appearing in a linear representation of $\{y_t\}$ that does not contain the Markov switching component.

Ghysels (2000b) discusses generalizations of periodic Markov switching models that feature higher-order AR dynamics and also include multivariate processes. Ghysels, McCulloch, and Tsay (1998) and Bac, Chevet, and Ghysels (1999) provide empirical examples involving U.S. aggregate macroeconomic time series as well as historical wheat price series.

Epilogue

The subject of seasonality has evolved a great deal over the past decade. As noted in the Preface, several books and special journal issues have been devoted to the subject in recent years. This monograph has attempted to review the principal theoretical developments in the field that have taken place since the work of Hylleberg (1986), which provided a comprehensive treatment of seasonality and econometric modeling up until the mid-1980s. It is acknowledged that empirical studies have been largely neglected in this monograph. This is not because they are unimportant. Rather it is because there are important complications to dealing with real data and a superficial treatment of these issues would not do them justice. We hope that other authors will take up the challenge of providing a comprehensive empirical treatment of seasonality in economic and financial time series.

The 1980s and 1990s have seen giant leaps forward made in asymptotic distribution theory for linear nonstationary stochastic processes. Although attention has obviously focused on zero-frequency behavior and associated tests for unit root nonstationarity, this has also led to windfall advances for seasonal time processes with possible seasonal unit roots. There is now a large degree of consensus about zero-frequency unit root behavior. Almost all data are first differenced without any further discussion, since the unit root hypothesis is taken for granted. There are nevertheless some serious questions about the adequacy of the zero-frequency unit root representation lingering in the back of any sophisticated time series econometrician's mind. For instance, the unit root hypothesis may be accepted because of the presence of long memory or because structural breaks are ignored. The poor finite sample power properties and the potentially serious size distortions (particularly when moving average roots nearly cancel out autoregressive unit roots) are also causing concern. To bridge the gap between the properties of stationary and nonstationary stochastic processes, there is a literature on near-nonstationarity. All these arguments and developments apply in the seasonal case as well. Despite some reservations, it is clear that the unit root hypothesis is well established for the zero frequency. There is, however, less concensus on the existence of seasonal unit roots.

Dealing with seasonal nonstationarity is typically done by one of two ways. Either first differences are assumed to follow a deterministic seasonal process, or seasonal differences are assumed to be covariance stationary. For the purpose of short-term forecasting the difference may not matter a great deal, in a similar way that the difference between a linear trend and a unit root process does not generally have much of an impact on short-term forecasting in a nonseasonal context. Nevertheless, the distinctive theoretical implications of these two types of processes are important for the analysis of longrun properties, including cointegration. The inadequacies of standard tests for seasonal unit roots are severe, as noted in Chapters 2 and 3. Samples are certainly small, and the near cancellation of MA and AR roots is usually a serious source of size distortions. Most seasonal adjustment procedures assume that seasonal differencing is the appropriate transformation. This is rather unfortunate and is often at odds with what applied econometricians do, since they often rely on dummy variables for seasonal variation. Despite the advances made in the analysis of linear seasonal processes, the picture at the moment is a rather complicated one for the applied worker. Appropriate tools of analysis are available, but there may be too little empirical information available on the nature of seasonality for it to be possible to give a correspondingly simple message to that available for the zero-frequency case. It is, however, possible that a serious attempt to develop economic theories of decision making in a seasonal context would shed more light on these issues.

Despite the present incomplete knowledge of the nature of seasonal nonstationarity, needless to say we are much in favor of using unadjusted data. The arguments discussed in Chapter 5 showing the dangers of filtering and estimation should be quite convincing. Yet, it is still a common practice, more so in North America than in Europe, to use seasonally adjusted series. The technology of the Web and the ease of data transfers and storage will probably lead to a situation in which raw data are made public more often in the future, and others will apply their own adjustment procedures. We have made considerable progress over the past decade in this regard. The X-11 program was described in an obscure government document that did not have wide circulation. Now, the X-12 program is clearly described in an academic journal and the program is downloadable from the Web. The leap forward in the transparency of adjustment procedures should be applauded. There is no doubt that seasonal adjustment will remain standard to enable the ready interpretation of prices and other published economic data for the public at large. Econometric practice could be improved upon by at least paying more attention to the potentially serious consequences of adjustment procedures.

Seasonality is an important, and largely predictable, aspect of shortrun movements in economic time series. One fascinating topic is how these seasonal movements interact with other types of fluctuations. Periodic processes give one angle on such an interaction, with seasonality and unit root behavior being inextricably intertwined in the case of periodically integrated processes. Another type of possible interaction is that between business cycles and seasonality. For example, models have been developed in which the probability of a regime switch from expansion to recession, or vice versa, is periodically varying. Despite this, and despite the recent explosion of interest in nonlinear business cycle modeling, the link between seasonal movements and business cycle regimes remains to date largely unexplored.

We expect that in the next decade there will be more research in the area of financial high-frequency data. This area is still in its infancy and hence very much unsettled. The nature of tick-by-tick data will prompt us to think about many new issues. For example, they will take us away from the traditional topics of testing for unit roots at seasonal frequencies. They will, however, also bring to the surface some old debates about seasonality and, in particular, seasonal adjustment. If a researcher wants to fit a structural model of micromarket behavior, then (s)he will wish to use the raw data and hence will face many questions about the appropriate treatment of seasonality that have been posed previously in the context of modeling economic time series. On the other hand, if high-frequency data are used simply to summarize certain trends, it may be advisable to apply prefiltering through some form of weekly or daily seasonal adjustment. High-frequency data may also provide more fertile ground for applications of periodic models. So far the models discussed in Chapter 6 have had only limited practical application because they are easily overparameterized and hence often not parsimonious when fitted to quarterly or monthly macroeconomic data. The need of periodic models may be more warranted with high-frequency data that feature complex intradaily repetitive dynamics. The sheer number of data points make it also possible to look at more richly parameterized structures. We hope and expect that the next comprehensive book on seasonality will pick up these many challenges.

Bibliography

Abeysinghe, T. (1991), "Inappropriate Use of Seasonal Dummies in Regression," *Economics Letters* **36**, 175–179.

Abeysinghe, T. (1994), "Deterministic Seasonal Models and Spurious Regression," *Journal of Econometrics* **61**, 259–272.

Ahtola, J. and G. C. Tiao (1987), "Distributions of Least Squares Estimators of Autoregressive Parameters for a Process with Complex Roots on the Unit Circle," *Journal of Time Series Analysis* **8**, 1–14.

Albert, J. and S. Chib (1993), "Bayesian Inference via Gibbs Sampling of Autoregressive Time Series Subject to Markov Measured Variance Shifts," *Journal of Business and Economic Statistics* **11**, 1–15.

Andel, J. (1976), "Autoregressive Series with Random Parameters," *Mathematische Operationsforschung und Statistik, Series Statistics* **7**, 736–741.

Andersen, T. G. and T. Bollerslev (1997a), "Heterogeneous Information Arrivals and Return Volatility Dynamics: Uncovering the Long-Run in High Frequency Returns," *Journal of Finance* **52**, 975–1005.

Andersen, T. G. and T. Bollerslev (1997b), "Intraday Periodicity and Volatility Persistence in Financial Markets," *Journal of Empirical Finance* **4**, 115–158.

Ansley, C. F. and P. Newbold (1980), "Finite Sample Properties of Estimators for Autoregressive Moving Average Models," *Journal of Econometrics* **13**, 159–183.

Bac, C., J. M. Chevet, and E. Ghysels (1999), "A Time Series Model with Periodic Stochastic Regime Switching, Part II: Applications to 16th and 17th Century Grain Prices," *Macroeconomic Dynamics* (forthcoming).

Baillie, R. T. and T. Bollerslev (1989), "The Message in Daily Exchange Rates: A Conditional Variance Tale," *Journal of Business and Economic Statistics* **7**, 297–305.

Baillie, R. T. and T. Bollerslev (1991), "Intra Day and Inter Market Volatility in Foreign Exchange Rates," *Review of Economic Studies* **58**, 565–585.

Baillie, R. T., T. Bollerslev, and H. O. Mikkelsen (1996), "Fractional Integrated Generalized Autoregressive Conditional Heteroskedasticity," *Journal of Econometrics* **74**, 3–30.

Banerjee, A., R. L. Lumsdaine, and J. H. Stock (1992), "Recursive and Sequential Tests of the Unit-Root and Trend-Break Hypotheses: Theory and International Evidence," *Journal of Business and Economic Statistics* **10**, 271–288.

Banerjee, A., J. Dolado, J. W. Galbraith, and D. F. Hendry (1993), *Co-Integration, Error-Correction and the Econometric Analysis of Non-Stationary Time Series*, Oxford: Oxford University Press.

Baxter, M. and R. G. King (1999), "Measuring Business Cycles: Approximate Band Pass Filters for Economic Time Series," *Review of Economics and Statistics* **81**, 575–593.

Beaulieu, J. J. and J. A. Miron (1993), "Seasonal Unit Roots in Aggregate U.S. Data," *Journal of Econometrics* **55**, 305–328.

Bell, W. R. (1984), "Signal Extraction for Nonstationary Time Series," *Annals of Statistics* **12**, 646–664.

Bell, W. R. (1987), "A Note on Overdifferencing and the Equivalence of Seasonal Time Series Models With Monthly Means and Models With $(0, 1, 1)_{12}$ Seasonal Parts When $\Theta = 1$," *Journal of Business and Economic Statistics* **5**, 383–387.

Bell, W. R. (1992), "On Some Properties of X-11 Symmetric Linear Filters," unpublished document, Statistical Research Division, U.S. Bureau of the Census.

Bell, W. R. and S. C. Hillmer (1984), "Issues Involved with the Seasonal Adjustment of Economic Time Series," *Journal of Business and Economic Statistics* **2**, 291–320.

Bell, W. R. and M. Kramer (1999), "Toward Variances for X-11 Seasonal Adjustments," *Survey Methodology* **25**, 13–29.

Bera, A. K. and M. L. Higgins (1993), "ARCH Models: Properties, Estimation, and Testing," *Journal of Economic Surveys* **7**, 305–366.

Birchenhall, C. R., R. C. Bladen-Hovell, A. P. L. Chui, D. R. Osborn, and J. P. Smith (1989), "A Seasonal Model of Consumption," *Economic Journal* **99**, 837–843.

Bobbitt, L. and M. C. Otto (1990), "Effects of Forecasts on the Revisions of Seasonally Adjusted Values Using the X-11 Seasonal Adjustment Procedure," *Proceedings of the Business and Economic Statistics Section*, Alexandria: American Statistical Association, pp. 449–453.

Bollerslev, T. (1986), "Generalized Autoregressive Conditional Heteroskedasticity," *Journal of Econometrics* **31**, 307–327.

Bollerslev, T., R. Y. Chou, and K. F. Kroner (1992), "ARCH Modeling in Finance: A Review of Theory and Empirical Evidence," *Journal of Econometrics* **52**, 5–59.

Bollerslev, T. and I. Domowitz (1993), "Trading Patterns and Prices in the Interbank Foreign Exchange Market," *Journal of Finance* **48**, 1421–1443.

Bollerslev, T., R. F. Engle, and D. B. Nelson (1994), "ARCH Models," in R. F. Engle and D. McFadden (eds.), *Handbook of Econometrics*, Amsterdam: Elsevier Science/North-Holland, Vol. 4, pp. 2959–3038.

Bollerslev, T. and E. Ghysels (1996), "Periodic Autoregressive Conditional Heteroskedasticity," *Journal of Business and Economic Statistics* **14**, 139–152.

Bollerslev, T. and R. J. Hodrick (1995), "Financial Market Efficiency Tests," in M. H. Pesaran and M. Wickens (eds.), *Handbook of Applied Econometrics: Macroeconomics*, Oxford Blackwell, pp. 415–458.

Bollerslev, T. and J. M. Wooldridge (1992), "Quasi Maximum Likelihood Estimation and Inference in Dynamic Models with Time Varying Covariances," *Econometric Reviews* **11**, 143–172.

Boswijk, H. P. (1994), "Testing for an Unstable Root in Conditional and Structural Error Correction Models," *Journal of Econometrics* **63**, 37–60.

Boswijk, H. P. and P. H. Franses (1995), "Periodic Cointegration: Representation and Inference," *Review of Economics and Statistics* **77**, 436–454.

Boswijk, H. P. and P. H. Franses (1996), "Unit Roots in Periodic Autoregressions," *Journal of Time Series Analysis* **17**, 221–245.

Bougerol, P. and N. Picard (1991), "Stationarity of GARCH Processes," *Journal of Econometrics* **52**, 115–127.

Box, G. E. P. and G. Jenkins (1976), *Time Series Analysis: Forecasting and Control*, Englewood Cliffs, NJ: Prentice-Hall, rev. ed.

Box, G. E. P. and G. C. Tiao (1975), "Intervention Analysis with Applications to Economic and Environmental Problems," *Journal of the American Statistical Association* **70**, 70–79.

Breitung, J. and P. H. Franses (1998), "On Phillips–Perron Type Tests for Seasonal Unit Roots," *Econometric Theory* **14**, 200–221.

Burridge, P. and A. M. R. Taylor (2000), "On the Properties of Regression-Based Tests for Seasonal Unit Roots in the Presence of Higher Order Serial Correlation," mimeo, City University, London.

Burridge, P. and K. F. Wallis (1984), "Unobserved Component Models for Seasonal Adjustment Filters," *Journal of Business and Economic Statistics* **2**, 350–359.

Burridge, P. and K. F. Wallis (1988), "Prediction Theory for Autoregressive Moving Average Processes," *Econometric Reviews* **7**, 65–95.

Burridge, P. and K. F. Wallis (1990), "Seasonal Adjustment and Kalman Filtering: Extension to Periodic Variances," *Journal of Forecasting* **9**, 109–118.

Caner, M. (1998), "A Locally Optimal Seasonal Unit-Root Test," *Journal of Business and Economic Statistics* **16**, 349–356.

Canova, F. and E. Ghysels (1994), "Changes in Seasonal Patterns: Are They Cyclical?," *Journal of Economic Dynamics and Control* **18**, 1143–1171.

Canova, F. and B. E. Hansen (1995), "Are Seasonal Patterns Constant Over Time? A Test for Seasonal Stability," *Journal of Business and Economic Statistics* **13**, 237–252.

Cecchetti, S. G. and P. Lam (1992), "What Do We Learn from Variance Ratio Statistics? A Study of Stationary and Nonstationary Models with Breaking Trends," mimeo, Ohio State University.

Chan, K., K. C. Chan, and G. A. Karolyi (1991), "Intraday Volatility in the Stock Index and Stock Index Futures Market," *Review of Financial Studies* **4**, 657–684.

Chan, N. H. (1989a), "On the Nearly Nonstationary Seasonal Time Series," *The Canadian Journal of Statistics* **17**, 279–284.

Chan, N. H. (1989b), "Asymptotic Inference for Unstable Autoregressive Time Series with Drifts," *Journal of Statistical Inference and Planning* **23**, 301–312.

Chan, N. H. and C. Z. Wei (1988), "Limiting Distributions of Least Squares Estimates of Unstable Autoregressive Processes," *Annals of Statistics* **16**, 367–401.

Chang, I. and G. C. Tiao (1983), "Estimation of Time Series Parameters in the Presence of Outliers," Technical Report 8, Statistics Research Center, University of Chicago.

Chesher, A. (1984), "Testing for Neglected Heterogeneity," *Econometrica* **52**, 865–872.

Cleveland, W. P. (1972), "Analysis and Forecasting of Seasonal Time Series," Ph.D. dissertation, Statistics Department, University of Wisconsin.

Cleveland, W. P. and G. C. Tiao (1976), "Decomposition of Seasonal Time Series: A Model for the X-11 Program," *Journal of the American Statistical Association* **71**, 581–587.

Cox, D. R. (1983), "Some Remarks on Overdispersion," *Biometrika* **70**, 269–274.

Dacorogna, M. M., U. A. Müller, R. J. Nagler, R. B. Olsen, and O. V. Pictet (1993), "A Geographical Model for the Daily and Weekly Seasonal Volatility in the Foreign Exchange Market," *Journal of International Money and Finance* **12**, 413–438.

Dagum, E. B. (1980), "The X-11-ARIMA Seasonal Adjustment Method," Report 12-564E, Statistics Canada, Ottawa.

Dagum, E. B. (1982), "The Effects of Asymmetric Filters on Seasonal Factor Revisions," *Journal of the American Statistical Association* **77**, 732–738.

Dagum, E. B. (1983), "Spectral Properties of the Concurrent and Forecasting Seasonal Linear Filters of the X-11 ARIMA Method," *The Canadian Journal of Statistics* **11**, 73–90.

da Silva Lopes, A. C. B. and A. Montanes (1999), "The Behaviour of Seasonal Unit Roots Tests with Seasonal Mean Shifts," Discussion Paper 10-99, Universidade Técnica de Lisboa.

De Gooijer, J. G. and P. H. Franses (1997), "Forecasting and Seasonality: Editors' Introduction to Special Issue," *International Journal of Forecasting* **13**, 303–305.

Deistler, M., W. Dunsmuir, and E. J. Hannan (1978), "Vector Linear Time Series Models: Corrections and Extensions," *Advances in Applied Probability* **8**, 360–372.

de Jong, P. (1991), "The Diffuse Kalman Filter," *Annals of Statistics* **19**, 1073–1083.

de Jong, P. and C.-C. Lin (1994), "Stationary and Nonstationary State Space Models," *Journal of Time Series Analysis* **15**, 151–166.

Dickey, D. A. (1993), "Discussion: Seasonal Unit Roots in Aggregate U.S. Data," *Journal of Econometrics* **55**, 329–331.

Dickey, D. A. and W. A. Fuller (1979), "Distribution of the Estimators for Autoregressive Time Series with a Unit Root," *Journal of the American Statistical Association* **74**, 427–431.

Dickey, D. A. and W. A. Fuller (1981), "Likelihood Ratio Statistics for Autoregressive Time Series with a Unit Root," *Econometrica* **49**, 1057–1072.

Dickey, D. A., D. P. Hasza, and W. A. Fuller (1984), "Testing for Unit Roots in Seasonal Time Series," *Journal of the American Statistical Association* **79**, 355–367.

Dickey, D. A. and S. G. Pantula (1987), "Determining the Order of Differencing in Autoregressive Processes," *Journal of Business and Economic Statistics* **5**, 455–461.

Diebold, F. X., J.-H. Lee, and G. C. Weinbach (1994), "Regime Switching with Time-Varying Transition Probabilities" in C. Hargreaves (ed.), *Nonstationarity Time Series Analysis and Cointegration*, Oxford: Oxford University Press.

Drost, F. C. and T. E. Nijman (1993), "Temporal Aggregation of GARCH Processes," *Econometrica* **61**, 909–928.

Dunsmuir, W. and E. J. Hannan (1976), "Vector Linear Time Series Models," *Advances in Applied Probability* **8**, 339–364.

Durland, M. J. and T. H. McCurdy (1994), "Duration-Dependent Transitions in a Markov Model of U.S. GNP Growth," *Journal of Business and Economic Statistics* **22**, 279–288.

Engle, R. F. (1982), "Autoregressive Conditional Heteroskedasticity with Estimates of the Variance of U.K. Inflation," *Econometrica* **50**, 987–1008.

Engle, R. F. (1990), "Discussion: Stock Market Volatility and the Crash of 87," *Review of Financial Studies* **3**, 103–106.

Engle, R. F. (2000), "The Econometrics of Ultra-High Frequency Data," *Econometrica* **68**, 1–22.

Engle, R. F. and C. W. J. Granger (1987), "Cointegration and Error Correction: Representation, Estimation and Testing," *Econometrica* **55**, 251–276.

Engle, R. F., C. W. J. Granger, and J. J. Hallman (1988), "Merging Short- and Long-Run Forecasts: An Application of Seasonal Cointegration to Monthly Electricity Sales Forecasting," *Journal of Econometrics* **40**, 45–62.

Engle, R. F., C. W. J. Granger, S. Hylleberg, and H. S. Lee (1993), "Seasonal Cointegration: The Japanese Consumption Function," *Journal of Econometrics* **55**, 275–303.

Engle, R. F., T. Ito, and W.-L. Lin (1990), "Meteor Showers or Heat Waves? Heteroskedastic Intra-Daily Volatility in the Foreign Exchange Market," *Econometrica* **58**, 525–542.

Engle, R. and J. Russell (1997), "Forecasting the Frequency of Changes in Quoted Foreign Exchange Prices with the Autoregressive Conditional Duration Model," *Journal of Empirical Finance* **4**, 187–212.

Engle, R. and J. Russell (1998), "Autoregressive Conditional Multinomial: A New Model for Irregularly Spaced Discrete-Valued Time Series Data with Applications to High Frequency Financial Data," *Econometrica* **66**, 1127–1162.

Ericsson, W. R., D. F. Hendry, and H.-A. Tran (1994), "Cointegration Seasonality, Encompassing and the Demand for Money in the United Kingdom" in C. P. Hargreaves (ed.), *Nonstationary Time Series Analysis and Cointegration,* Oxford: Oxford University Press, 179–224.

Filardo, A. J. (1994), "Business Cycle Phases and Their Transitional Dynamics," *Journal of Business and Economic Statistics* **12**, 299–308.

Findley, D. F., B. C. Monsell, W. R. Bell, M. C. Otto, and B.-C. Chen (1998), "New Capabilities and Methods of the X-12-ARIMA Seasonal-Adjustment Program," *Journal of Business and Economic Statistics* **16**, 127–177 (with discussion).

Findley, D. F., B. C. Monsell, H. B. Shulman, and M. G. Pugh (1990), "Sliding Spans Diagnostics for Seasonal and Related Adjustments," *Journal of the American Statistical Association* **85**, 345–355.

Fiorentini, G. and A. Maravall (1996), "Unobserved Components in ARCH Models," *Journal of Forecasting* **15**, 175–201.

Flores, R. and A. Novales (1997), "A General Test for Univariate Seasonality," *Journal of Time Series Analysis* **18**, 29–48.

Foster, F. D. and S. Viswanathan (1995), "Can Speculative Trading Explain the Volume-Volatility Relation?" *Journal of Business and Economic Statistics* **13**, 379–396.

Franses, P. H. (1991), "Seasonality, Nonstationarity and the Forecasting of Monthly Time Series," *International Journal of Forecasting* **7**, 199–208.

Franses, P. H. (1993), "A Method to Select Between Periodic Cointegration and Seasonal Cointegration," *Economics Letters* **41**, 7–10.

Franses, P. H. (1994), "A Multivariate Approach to Modeling Univariate Seasonal Time Series," *Journal of Econometrics* **63**, 133–151.

Franses, P. H. (1995a), "A Vector of Quarters Representation for Bivariate Time Series," *Econometric Reviews* **14**, 55–63.

Franses, P. H. (1995b), "The Effects of Seasonally Adjusting a Periodic Autoregressive Process," *Computational Statistics and Data Analysis* **19**, 683–704.

Franses, P. H. (1996a), "Recent Advances in Modelling Seasonality," *Journal of Economic Surveys* **10**, 299–345.

Franses, P. H. (1996b), *Periodicity and Stochastic Trends in Economic Time Series*, Oxford: Oxford University Press.

Franses, P. H., S. Hylleberg, and H. S. Lee (1995), "Spurious Deterministic Seasonality," *Economics Letters* **48**, 241–248.

Franses, P. H. and T. Kloek (1995), "A Periodic Cointegration Model of Quarterly Consumption," *Applied Stochastic Models and Data Analysis* **11**, 159–166.

Franses, P. H. and R. Kunst (1999), "On the Role of Seasonal Intercepts in Seasonal Cointegration," *Oxford Bulletin of Economics and Statistics* **61**, 409–434.

Franses, P. H. and M. McAleer (1998), "Cointegration Analysis of Seasonal Time Series," *Journal of Economic Surveys* **12**, 651–678.

Franses, P. H. and R. Paap (1994), "Model Selection in Periodic Autoregressions," *Oxford Bulletin of Economics and Statistics* **56**, 421–439.

Franses, P. H. and R. Paap (1995b), "Seasonality and Stochastic Trends in German Consumption and Income," *Empirical Economics* **20**, 109–132.

Franses, P. H. and G. Romijn (1993), "Periodic Integration in Quarterly UK Macroeconomic Variables," *International Journal of Forecasting* **9**, 467–476.

Franses, P. H. and T. Vogelsang (1998), "On Seasonal Cycles, Unit Roots, and Mean Shifts," *Review of Economics and Statistics* **80**, 231–240.

French, K. R. and R. Roll (1986), "Stock Return Variances: The Arrival of Information and the Reaction of Traders," *Journal of Financial Economics* **17**, 5–26.

Fuller, W. A. (1996), *Introduction to Statistical Time Series*, New York: Wiley, 2nd ed.

Gallant, A. R., P. E. Rossi, and G. Tauchen (1992), "Stock Prices and Volume," *Review of Financial Studies* **5**, 199–242.

Gersovitz, M. and J. G. MacKinnon (1978), "Seasonality in Regression: An Application of Smoothness Priors," *Journal of the American Statistical Association* **73**, 264–273.

Geweke, J. (1978), "The Temporal and Sectoral Aggregation of Seasonally Adjusted Time Series" in A. Zellner (ed.), *Seasonal Analysis of Economic Time Series*, Washington, DC: U.S. Department of Commerce, Census Bureau.

Ghysels, E. (1984), "The Economics of Seasonality: The Case of the Money Supply," Unpublished Ph.D. dissertation, Department of Managerial Economics and Decision Science, Northwestern University.

Ghysels, E. (1988), "A Study Towards a Dynamic Theory of Seasonality for Economic Time Series," *Journal of the American Statistical Association* **83**, 168–172.

Ghysels, E. (1991), "Are Business Cycle Turning Points Uniformly Distributed Throughout the Year?" Discussion Paper 3891, C.R.D.E., Université de Montréal.

Ghysels, E. (1993), "On Scoring Asymmetric Periodic Probability models of Turning Point Forecasts," *Journal of Forecasting* **12**, 227–238.

Ghysels, E. (1994a), "On the Economics and Econometrics of Seasonality," in C. A. Sims (ed.), *Advances in Econometrics – Sixth World Congress of the Econometric Society*, Cambridge: Cambridge University Press, Vol. 1, Chap. 7, pp. 257–316.

Ghysels, E. (1994b), "On the Periodic Structure of the Business Cycle," *Journal of Business and Economic Statistics* **12**, 289–298.

Ghysels, E. (1997a), "Seasonal Adjustment and Other Data Transformations," *Journal of Business and Economic Statistics* **15**, 410–418.

Ghysels, E. (1997b), "On Seasonality and Business Cycle Durations: A Nonparametric Investigation," *Journal of Econometrics* **79**, 269–290.

Ghysels, E. (2000a), "Some Econometric Recipes for High Frequency Data Cooking," *Journal of Business and Economic Statistics* **18**, 154–163.

Ghysels, E. (2000b), "A Time Series Model with Periodic Stochastic Regime Switching, Part I: Theory," *Macroeconomic Dynamics* (forthcoming).

Ghysels, E., C. Gouriéroux, and J. Jasiak (1997), "High Frequency Financial Time Series Data: Some Stylized Facts and Models of Stochastic Volatility," in C. Dunis and B. Zhou (eds.), *Nonlinear Modelling of High Frequency Financial Time Series*, New York: Wiley, Chap. 7, pp. 127–159.

Ghysels, E., C. Gouriéroux, and J. Jasiak (1998a), "Testing Causality in Financial Market Transition Dynamics: An Application to Return, Spreads and Volume," Discussion Paper, Centre de Recherche en Economie et Statistique, Paris.

Ghysels, E., C. Gouriéroux, and J. Jasiak (1998b), "Stochastic Volatility Durations," Discussion Paper, Centre de Recherche en Economie et Statistique, Paris.

Ghysels, E., C. W. J. Granger, and P. Siklos (1996), "Is Seasonal Adjustment a Linear or Nonlinear Data Filtering Process?" *Journal of Business and Economic Statistics* **14**, 374–386 (with discussion).

Ghysels, E., C. W. J. Granger, and P. Siklos (1998), "Seasonal Adjustment and Volatility Dynamics," Working Paper, Penn State University.

Ghysels, E., A. Hall, and H. S. Lee (1996), "On Periodic Structures and Testing for Seasonal Unit Roots," *Journal of the American Statistical Association* **91**, 1551–1559.

Ghysels, E., H. S. Lee, and J. Noh (1994), "Testing for Unit Roots in Seasonal Time Series – Some Theoretical Extensions and a Monte Carlo Investigation," *Journal of Econometrics* **62**, 415–442.

Ghysels, E. and O. Lieberman (1996), "Dynamic Regression and Filtered Data Series: A Laplace Approximation to the Effects of Filtering in Small Samples," *Econometric Theory* **12**, 432–457.

Ghysels, E., R. McCulloch, and R. Tsay (1998), "Bayesian Inference for a General Class of Periodic Markov Switching Regime Models," *Journal of Applied Econometrics* **13**, 129–144.

Ghysels, E. and P. Perron (1993), "The Effect of Seasonal Adjustment Filters on Tests for a Unit Root," *Journal of Econometrics* **55**, 57–99.

Gibbons, M. R. and P. J. Hess (1981), "Day of the Week Effects and Asset Returns," *Journal of Business* **54**, 579–596.

Gladyshev, E. G. (1961), "Periodically Correlated Random Sequences," *Soviet Mathematics* **2**, 385–388.

Glosten, L. R., R. Jagannathan, and D. E. Runkle (1993), "On the Relation Between the Expected Value and the Volatility of the Nominal Excess Return on Stocks," *Journal of Finance* **48**, 1779–1801.

Godfrey, M. D. and H. Karreman (1967), "A Spectrum Analysis of Seasonal Adjustment," in M. Shubik (ed.), *Essays in Mathematical Economics in Honor of Oscar Morgenstern*, Princeton, NJ: Princeton University Press.

Goméz, V. (1999), "Three Equivalent Methods for Filtering Finite Nonstationary Time Series," *Journal of Business and Economic Statistics* **17**, 109–116.

Goméz, V. and A. Maravall (1994), "Estimation, Prediction and Interpolation for Nonstationary Series with the Kalman Filter," *Journal of the American Statistical Association* **86**, 611–624.

Goméz, V. and A. Maravall (1996), "Programs TRAMO and SEATS, Instructions for the User (Beta version: September 1996)," Working Paper 9628, Bank of Spain.

Goodhart, C. A. E., S. G. Hall, S. G. B. Henry, and B. Pesaran (1993), "New Effects in a High-Frequency Model of the Sterling-Dollar Exchange Rate," *Journal of Applied Econometrics* **8**, 1–13.

Gouriéroux, C., J. Jasiak, and G. LeFol (1999), "Intraday Market Activity," *Journal of Financial Markets* **3**, 193–226.

Granger, C. W. J. (1978), "Seasonality: Causation, Interpretation and Implications," in A. Zellner (ed.), *Seasonal Analysis of Economic Time Series*, Washington, DC: Department of Commerce.

Granger, C. W. J. (1986), "Developments in the Study of Cointegrated Economic Variables," *Oxford Bulletin of Economics and Statistics* **48**, 213–227.

Granger, C. W. J. and M. Hatanaka (1964), *Spectral Analysis of Economic Time Series*, Princeton, NJ: Princeton University Press.

Granger, C. W. J. and P. Newbold (1974), "Spurious Regressions in Econometrics," *Journal of Econometrics* **2**, 111–120.

Granger, C. W. J. and N. R. Swanson (1997), "An Introduction to Stochastic Unit Roots," *Journal of Econometrics* **80**, 35–62.

Hall, A. R. (1994), "Testing for a Unit Root in Time Series With Pretest Data-Based Model Selection," *Journal of Business and Economic Statistics* **12**, 461–470.

Hamao, Y., R. W. Masulis, and V. Ng (1990), "Correlations in Price Changes and Volatility Across International Stock Markets," *Review of Financial Studies* **3**, 281–307.

Hamilton, J. D. (1988), "Rational-Expectations Econometric Analysis of Changes in Regime: An Investigation of the Term Structure of Interest Rates," *Journal of Economic Dynamics and Control* **12**, 385–423.

Hamilton, J. D. (1989), "A New Approach to the Economic Analysis of Nonstationary Time Series and the Business Cycle," *Econometrica* **57**, 357–384.

Hamilton, J. D. (1990), "Analysis of Time Series Subject to Changes in Regime," *Journal of Econometrics* **45**, 39–70.

Hamilton, J. D. (1994a), *Time Series Analysis*, Princeton, NJ: Princeton University Press.

Hamilton, J. D. (1994b), "State-Space Models" in R. F. Engle and D. McFadden (eds.), *Handbook of Econometrics 4*, Amsterdam: North-Holland, pp. 3038–3080.

Hannan, E. J. (1963), "The Estimation of the Seasonal Variation in Economic Time Series," *Journal of the American Statistical Association* **58**, 31–44.

Hannan, E. J. (1967), "Measurement of a Wandering Signal Amid Noise," *Journal of Applied Probability* **4**, 90–102.

Hannan, E. J. (1973), "The Asymptotic Theory of Linear Time Series Models," *Journal of Applied Probability* **10**, 130–145.

Hannan, E. J. and J. Rissanen (1982), "Recursive Estimation of Mixed Autoregressive-Moving Average Order," *Biometrika* **69**, 81–94.

Hannan, E. J., R. D. Terrell, and N. E. Tuckwell (1970), "The Seasonal Adjustment of Economic Time Series," *International Economic Review* **11**, 24–52.

Hansen, L. P. and T. J. Sargent (1993), "Seasonality and Approximation Errors in Rational Expectations Models," *Journal of Econometrics* **55**, 21–56.

Harvey, A. C. (1989), *Forecasting Structural Time Series Models and the Kalman Filter*, Cambridge: Cambridge University Press.

Harvey, A. C. (1993), *Time Series Models*, Cambridge, MA: MIT Press, 2nd ed.

Harvey, A. C. and R. G. Pierse (1984), "Estimating Missing Observations in Economic Time Series," *Journal of the American Statistical Association* **79**, 125–131.

Harvey, A. C. and P. M. J. Todd (1983), "Forecasting Economic Time Series with Structural and Box-Jenkins Models," *Journal of Business and Economic Statistics* **1**, 299–315 (with discussion).

Harvey, C. R. and R. D. Huang (1991), "Volatility in the Foreign Currency Futures Market," *Review of Financial Studies* **4**, 543–569.

Hasza, D. P. and W. A. Fuller (1982), "Testing for Nonstationarity Parameter Specifications in Seasonal Time Series Models," *Annals of Statistics* **10**, 209–216.

Herwartz, H. (1997), "Performance of Periodic Error Correction Models in Forecasting Consumption Data," *International Journal of Forecasting* **13**, 421–431.

Hillmer, S. C. and G. C. Tiao (1982), "An ARIMA-Model Based Approach to Seasonal Adjustment," *Journal of the American Statistical Association* **77**, 63–70.

Hodrick, R. J. and E. C. Prescott (1997), "Postwar U.S. Business Cycles: An Empirical Investigation," *Journal of Money, Credit and Banking* **29**, 1–16.

Hylleberg, S. (1986), *Seasonality in Regression*, New York: Academic Press.

Hylleberg, S. (ed.) (1992), *Modelling Seasonality*, Oxford: Oxford University Press.

Hylleberg, S. (1994), "Modelling Seasonal Variation," in C. P. Hargreaves (ed.), *Nonstationary Time Series Analysis and Cointegration*, Oxford: Oxford University Press.

Hylleberg, S. (1995), "Tests for Seasonal Unit Roots: General to Specific or Specific to General?," *Journal of Econometrics* **69**, 5–25.

Hylleberg, S., R. F. Engle, C. W. J. Granger, and B. S. Yoo (1990), "Seasonal Integration and Cointegration," *Journal of Econometrics* **44**, 215–238.

Hylleberg, S., C. Jørgensen, and N. K. Sørensen (1993), "Seasonality in Macroeconomic Time Series," *Empirical Economics* **18**, 321–335.

Ilmakunnas, P. (1990), "Testing the Order of Differencing in Quarterly Data: An Illustration of the Testing Sequence," *Oxford Bulletin of Economics and Statistics* **52**, 79–87.

Jaditz, T. (1996), "Seasonality in Variances is Common in Macro Time Series," Discussion Paper, U.S. Bureau of Statistics.

Johansen, S. (1988), "Statistical Analysis of Cointegration Vectors," *Journal of Economic Dynamics and Control* **12**, 231–254.

Johansen, S. (1991), "Estimation and Hypothesis Testing of Cointegrating Vectors in Gaussian Vector Autoregressive Models," *Econometrica* **59**, 1551–1580.

Johansen, S. (1994), "The Role of the Constant and Linear Terms in Cointegration Analysis of Nonstationary Variables," *Econometric Reviews* **13**, 205–229.

Johansen, S. and K. Juselius (1990), "Maximum Likelihood Estimation and Inference on Cointegration – With Applications to the Demand for Money," *Oxford Bulletin of Economics and Statistics* **52**, 169–210.

Johansen, S. and E. Schaumburg (1999), "Likelihood Analysis of Seasonal Cointegration," *Journal of Econometrics* **88**, 301–339.

Kim, C. J. (1994), "Dynamic Linear Models with Markov Switching," *Journal of Econometrics* **60**, 1–22.

Kiviet, J. F. and G. D. A. Phillips (1992), "Exact Similar Tests for Unit Roots and Cointegration," *Oxford Bulletin of Economics and Statistics* **54**, 349–367.

Kleibergen, F. R. and P. H. Franses (1999), "Cointegration in a Periodic Vector Autoregression," Technical Report EI-9906/A, Econometric Report, University of Rotterdam.

Kohn, R. and C. F. Ansley (1986), "Prediction Mean Squared Error for State Space Models with Estimated Parameters," *Biometrika* **73**, 467–473.

Koopman, S. J., A. C. Harvey, J. A. Doornik, and N. Shephard (1996), *Stamp: Structural Time Series Analyser, Modeller and Predictor*, London: Chapman & Hall.

Kullback, S. and R. A. Leibler (1951), "On Information and Sufficiency," *Annals of Mathematical Statistics* **22**, 79–86.

Kunst, R. M. (1993), "Seasonal Cointegration in Macroeconomic Systems: Case Studies for Small and Large European Countries," *Review of Economics and Statistics* **75**, 325–330.

Kunst, R. M. (1997), "Testing for Cyclical Non-Stationarity in Autoregressive Processes," *Journal of Time Series Analysis* **18**, 123–135.

Kunst, R. M. and P. H. Franses (1998), "The Impact of Seasonal Constants on Forecasting Seasonally Cointegrated Time Series," *Journal of Forecasting* **17**, 109–124.

Kwiatkowski, D., P. C. B. Phillips, P. Schmidt, and Y. Shin (1992), "Testing the Null Hypothesis of Stationarity against the Alternative of a Unit Root," *Journal of Econometrics*, **54**, 159–178.

Laroque, G. (1977), "Analyse d'une Méthode de Désaisonnalisaton: Le Programme X-11 du US Bureau of the Census Version Trimestrielle," *Annales de l'INSEE* **88**, 105–127.

Laux, P. and L. K. Ng (1993), "The Sources of GARCH: Empirical Evidence from an Intraday Returns Model Incorporating Systematic and Unique Risks," *Journal of International Money and Finance* **12**, 543–560.

Lee, H. S. (1992), "Maximum Likelihood Inference on Cointegration and Seasonal Cointegration," *Journal of Econometrics* **54**, 1–49.

Lee, S. W. and B. E. Hansen (1994), "Asymptotic Theory for the GARCH(1,1) Quasi-Maximum Likelihood Estimator," *Econometric Theory* **10**, 29–52.

Leybourne, S. J., B. P. M. McCabe, and T. C. Mills (1996), "Randomized Unit Root Processes for Modelling and Forecasting Financial Time Series: Theory and Applications," *Journal of Forecasting* **15**, 253–270.

Leybourne, S. J., B. P. M. McCabe, and A. R. Tremayne (1996), "Can Economic Time Series Be Differenced to Stationarity?" *Journal of Business and Economic Statistics* **14**, 435–446.

Locke, P. R. and C. L. Sayers (1993), "Intra-day Futures Price Volatility: Information Effects and Variance Persistence," *Journal of Applied Econometrics* **8**, 15–30.

Lumsdaine, R. L. (1996), "Consistency and Asymptotic Normality of the Quasi-Maximum Likelihood Estimator in IGARCH(1,1) and Covariance Stationary IGARCH(1,1) Models," *Econometrica* **64**, 575–596.

Lütkepohl, H. (1984), "Linear Transformations of Vector ARMA Processes," *Journal of Econometrics* **26**, 283–293.

Lütkepohl, H. (1991), *Introduction to Multiple Time Series Analysis*, Berlin: Springer-Verlag.

Macauley, F. R. (1931), *The Smoothing of Time Series*, New York: National Bureau of Economic Research.

MacKinnon, J. G. (1991), "Critical Values for Co-Integration Tests," in R. F. Engle and C. W. J. Granger (eds.), *Long-Run Economic Relationships,* Oxford: Oxford University Press.

Maravall, A. (1985), "On Structural Time Series Models and the Characterization of Components," *Journal of Business and Economic Statistics* **3**, 350–355.

Maravall, A. (1988), "A Note on Minimum Mean Squared Error Estimation of Signals with Unit Roots," *Journal of Economic Dynamics and Control* **12**, 589–593.

Maravall, A. (1995), "Unobserved Components in Economic Time Series," in H. Pesaran, P. Schmidt, and M. Wickens (eds.), *The Handbook of Applied Econometrics*, Oxford: Blackwell, Vol. 1.

McCabe, B. P. M. and A. R. Tremayne (1995), "Testing a Time Series for Difference Stationarity," *Annals of Statistics* **23**, 1015–1028.

McCulloch, R. E. and R. S. Tsay (1994), "Bayesian Analysis of Autoregressive Time Series via the Gibbs Sampler," *Journal of Time Series Analysis* **15**, 235–250.

Miron, J. A. (1994), "The Economics of Seasonal Cycles," in C. A. Sims (ed.), *Advances in Econometrics – Sixth World Congress*, Cambridge: Cambridge University Press, Vol. I, Chap. 6, pp. 213–251.

Miron, J. A. (1996), *The Economics of Seasonal Cycles*, Cambridge, MA: MIT Press.

Nankervis, J. C. and N. E. Savin (1985), "Testing the Autoregressive Parameter with the t-Statistics," *Journal of Econometrics* **27**, 143–161.

Nelson, D. B. (1990), "Stationarity and Persistence in the GARCH(1,1) Model," *Econometric Theory* **6**, 318–334.

Nelson, D. B. (1991), "Conditional Heteroskedasticity in Asset Returns: A New Approach," *Econometrica* **59**, 347–370.

Nelson, D. B. and C. Q. Cao (1992), "Inequality Constraints in the Univariate GARCH Model," *Journal of Business and Economic Statistics* **10**, 229–235.

Nerlove, M. (1964), "Spectral Analysis of Seasonal Adjustment Procedures," *Econometrica* **32**, 241–286.

Nerlove, M., D. M. Grether, and J. L. Carvalho (1995), *Analysis of Economic Time Series. A Synthesis*, New York: Academic Press, rev. ed.

Newbold, P., C. Agiakloglou, and J. Miller (1994), "Adventures with ARIMA Software," *International Journal of Forecasting* **10**, 573–581.

Newey, W. K. and K. D. West (1987), "A Simple, Positive, Semi-Definite, Heteroskedasticity and Autocorrelation Consistent Covariance Matrix," *Econometrica* **55**, 703–708.

Nijman, T. and E. Sentana (1996), "Marginalization and Contemporaneous Aggregation in Multivariate GARCH Processes," *Journal of Econometrics* **71**, 71–87.

Novales, A. and R. F. de Fruto (1997), "Forecasting with Periodic Models: A Comparison with Time Invariant Coefficient Models," *International Journal of Forecasting* **13**, 393–405.

Nyblom, J. (1989), "Testing for the Constancy of Parameters Over Time," *Journal of the American Statistical Association* **84**, 223–230.

Osborn, D. R. (1982), "On the Criteria Functions Used for the Estimation of Moving Average Processes," *Journal of the American Statistical Association* **77**, 388–392.

Osborn, D. R. (1988), "Seasonality and Habit Persistence in a Life-Cycle Model of Consumption," *Journal of Applied Econometrics* **3**, 255–266.

Osborn, D. R. (1991), "The Implications of Periodically Varying Coefficients for Seasonal Time-Series Processes," *Journal of Econometrics* **48**, 373–384.

Osborn, D. R. (1993), "Discussion: Seasonal Cointegration," *Journal of Econometrics* **55**, 299–303.

Osborn, D. R., A. P. L. Chui, J. P. Smith, and C. R. Birchenhall (1988), "Seasonality and the Order of Integration for Consumption," *Oxford Bulletin of Economics and Statistics* **50**, 361–377.

Osborn, D. R., S. Heravi, and C. R. Birchenhall (1999), "Seasonal Unit Roots and Forecasts of Two-Digit European Industrial Production," *International Journal of Forecasting* **15**, 27–47.

Osborn, D. R. and P. M. M. Rodrigues (1998), "Asymptotic Distributions of Seasonal Unit Root Tests: A Unifying Approach," Discussion Paper 9811, School of Economic Studies, University of Manchester.

Osborn, D. R. and J. P. Smith (1989), "The Performance of Periodic Autoregressive Models in Forecasting Seasonal U.K. Consumption," *Journal of Business and Economic Statistics* **7**, 117–127.

Paap, R. and P. H. Franses (1999), "On Trends and Constants in Periodic Autoregressions," *Econometric Reviews* **18**, 271–286.

Paap, R., P. H. Franses, and H. Hoek (1997), "Bayesian Analysis of Seasonal Unit Roots and Seasonal Mean Shifts," *Journal of Econometrics* **78**, 359–380.

Pagano, M. (1978), "On Periodic and Multiple Autoregressions," *Annals of Statistics* **6**, 1310–1317.

Perron, P. (1989), "The Great Crash, the Oil Price Shock and the Unit Root Hypothesis," *Econometrica* **57**, 1361–1401.

Perron, P. (1991), "The Limiting Distribution of the Least-Squares Estimator in Nearly Integrated Seasonal Models," *The Canadian Journal of Statistics* **20**, 121–134.

Perron, P. and T. Vogelsang (1992), "Nonstationarity and Level Shifts with an Application to Purchasing Power Parity," *Journal of Business and Economic Statistics* **10**, 301–320.

Pfeffermann, D. (1994), "A General Method for Estimating the Variances of X-11 Seasonally Adjusted Estimators," *Journal of Time Series Analysis* **15**, 85–116.

Pfeffermann, D., M. Morry, and P. Wong (1995), "Estimation of the Variances of X-11-ARIMA Seasonally Adjusted Estimators for a Multiplicative Decomposition and Heteroscedastic Variances," *International Journal of Forecasting* **11**, 271–283.

Phillips, K. L. (1991), "A Two-Country Model of Stochastic Output with Changes in Regime," *Journal of International Economics* **31**, 121–142.

Phillips, P. C. B. (1986), "Understanding Spurious Regressions in Econometrics," *Journal of Econometrics* **33**, 311–340.

Phillips, P. C. B. (1987), "Towards a Unified Asymptotic Theory for Autoregression," *Biometrika* **74**, 535–547.

Phillips, P. C. B. and P. Perron (1988), "Testing for a Unit Root in Time Series Regression," *Biometrika* **75**, 335–346.

Pierce, D. A. (1979), "Signal Extraction Error in Nonstationary Time Series," *Annals of Statistics* **7**, 1303–1320.

Priestley, M. B. (1981), *Spectral Analysis and Time Series*, London: Academic Press, Vol. 1.

Proietti, T. (1998), "Spurious Periodic Autoregressions," *Econometrics Journal* **1**, C1–C22.

Psaradakis, Z. (1997), "Testing for Unit Roots in Time Series with Nearly Deterministic Seasonal Variation," *Econometric Reviews* **16**, 421–439.

Racine, M. D. (1997), "Modelling Seasonality in U.S. Stock Market Returns," Discussion Paper, Wilfrid Laurier University.

Rodrigues, P. M. M. (1999), "Near Seasonal Integration," *Econometric Theory*, forthcoming.

Rodrigues, P. M. M. and D. R. Osborn (1999a), "Performance of Seasonal Unit Root Tests for Monthly Data," *Journal of Applied Statistics* **26**, 985–1004.

Rodrigues, P. M. M. and D. R. Osborn (1999b), "Asymptotic Confidence Intervals and Seasonal Unit Root Test Statistics," Discussion Paper 9916, School of Economic Studies, University of Manchester.

Rose, D. F. (1977), "Forecasting Aggregates of Independent ARIMA Processes," *Journal of Econometrics* **5**, 323–345.

Savin, N. E. (1984), "Multiple Hypothesis Tests," in Z. Griliches and M. D. Intriligator (eds.), *Handbook of Econometrics 2*, Amsterdam: North-Holland, pp. 827–879.

Schwert, G. W. (1989), "Tests for Unit Roots: A Monte Carlo Investigation," *Journal of Business and Economic Statistics* **7**, 147–159.

Schwert, G. W. (1990), "Indexes of U.S. Stock Prices from 1802 to 1987," *Journal of Business* **63**, 399–426.

Shiskin, J., A. H. Young, and J. C. Musgrave (1967), "The X-11 Variant of the Census Method II Seasonal Adjustment Program," Technical Paper 15, U.S. Department of Commerce, Bureau of Economic Analysis.

Sichel, D. E. (1994), "Inventories and the Three Phases of the Business Cycle," *Journal of Business and Economic Statistics* **12**, 269–278.

Sims, C. A. (1974), "Seasonality in Regression," *Journal of the American Statistical Association* **69**, 618–626.

Sims, C. A. (1993), "Rational Expectations Modelling with Seasonally Adjusted Data," *Journal of Econometrics* **55**, 9–20.

Smith, J. and J. Otero (1997), "Structural Breaks and Seasonal Integration," *Economics Letters* **56**, 13–19.

Smith, R. J. and A. M. R. Taylor (1998), "Additional Critical Values and Asymptotic Representations for Seasonal Unit Root Tests," *Journal of Econometrics* **85**, 269–288.

Smith, R. J. and A. M. R. Taylor (1999), "Likelihood Ratio Tests for Seasonal Unit Roots," *Journal of Time Series Analysis* **20**, 453–476.

Sobel, E. L. (1967), "Prediction of a Noise-Distorted Multivariate Non-Stationary Signal," *Journal of Applied Probability* **4**, 330–342.

Tam, W.-K. and G. C. Reinsel (1997), "Tests for Seasonal Unit Root in ARIMA Models," *Journal of the American Statistical Association* **92**, 725–738.

Tam, W.-K. and G. C. Reinsel (1998), "Seasonal Moving-Average Unit Root Tests in the Presence of a Linear Trend," *Journal of Time Series Analysis* **19**, 609–625.

Tanaka, K. (1996), *Time Series Analysis: Nonstationarity and Noninvertible Distribution Theory*, New York: Wiley.

Taylor, A. M. R. (1997), "On the Practical Problems of Computing Seasonal Unit Root Tests," *International Journal of Forecasting* **13**, 307–318.

Taylor, A. M. R. and R. J. Smith (1998), "Tests of the Seasonal Unit Root Hypothesis Against Heteroscedastic Seasonal Integration," *Journal of Business and Economic Statistics* (forthcoming).

Thomson, P. and T. Ozaki (1992), "Transformation and Seasonal Adjustment," Technical Report, Institute of Statistics and Operations Research, Victoria University.

Tiao, G. and M. Grupe (1980), "Autoregressive-Moving Average Models in Time Series Data," *Biometrika* **67**, 365–373.

Troutman, B. M. (1979), "Some Results of Periodic Autoregression," *Biometrika* **66**, 365–373.

Tsay, R. S. (1987), "Conditional Heteroscedastic Time Series Models," *Journal of the American Statistical Association* **82**, 590–604.

Wallis, K. F. (1974), "Seasonal Adjustment and Relations Between Variables," *Journal of the American Statistical Association* **69**, 18–32.

Wallis, K. F. (1977), "Multiple Time Series Analysis and the Final Form of Econometric Models," *Econometrica* **45**, 1481–1497.

Wei, W. W. S. (1978), "Some Consequences of Temporal Aggregation in Seasonal Time

Series Models" in A. Zellner (ed.), *Seasonal Analysis of Economic Time Series*, Economic Research Report ER-1, Washington, DC: Bureau of the Census.

Weiss, A. A. (1986), "Asymptotic Theory for ARCH Models: Estimation and Testing," *Econometric Theory* **2**, 107–131.

Wells, J. M. (1997), "Modelling Seasonal Patterns and Long-Run Trends in U.S. Time Series, *International Journal of Forecasting* **13**, 407–420.

Wood, R., T. McInish, and J. K. Ord (1985), "An Investigation of Transaction Data for NYSE Stocks," *Journal of Finance* **40**, 723–739.

Young, A. H. (1965), "Estimating Trading-Day Variation in Monthly Economic Time Series," Technical Report 12, Bureau of the Census, U.S. Department of Commerce.

Young, A. H. (1968), "Linear Approximation to the Census and BLS Seasonal Adjustment Methods." *Journal of the American Statistical Association* **63**, 445–471.

Zakoian, J. M. (1994), "Threshold Heteroskedasticity Models," *Journal of Economic Dynamics and Control* **18**, 931–956.

Zivot, E. and D. W. K. Andrews (1992), "Further Evidence on the Great Crash, the Oil Price Shock and the Unit Root Hypothesis," *Journal of Business and Economic Statistics* **10**, 251–270.

Subject Index

Author Index